The Sapient Salesman

The Sapient Salesman

Spinning Life into Lessons,
One Tale at a Time

ERIN WILSON

Printed by CreateSpace
www.CreateSpace.com/5029365

Illustrations and back cover art by: Rosie Diaz

ISBN 978-0-9962373-8-3

SapientSalesman.com

TO THOSE WHO LET ME LAUGH OUT LOUD.

Table of Tales

Preface

The Sapient Salesman began as a blog and — while it has evolved from a humble internal sales blast to the well visited website it is today — it remains an outlet for my mishaps and musings alike. Now a book, *The Sapient Salesman: Spinning Life into Lessons, One Tale at a Time* is a collection of short stories, based on real-life events, that showcase the salesmanship — or lack there of — present in every-day interactions.

Since our shared experiences exhilarate us more than pedantic seminars, sales boot camps, and the next-best methodology — this collection faithfully showcases the humor, the hiccups, and hurdles we face daily. Each of these new and retold stories use otherwise ordinary events to bring into question the sales tactics and interpersonal philosophies we employ and [perhaps] take for granted. Through the introspection they inspire, you will discover opportunities to improve your own sales practices — both personally and professionally.

With a light spirit and a humble heart we just might find that there's a sapient salesman in all of us.

Disclaimer

During the last several years I've worked for, and with, several software companies and along the way have met many interesting people. While some may be mentioned by their natural nouns, most of the names of *The Sapient's* cast — my friends, from single serving to serious — have been changed to further the book's fictitious nature and provide me sufficient literary license.

So, let me just say this...

In no way do the thoughts or opinions represented here reflect those of any of the companies or parties portrayed in this book. Moreover, if it was funnier in my head than in real life, you got the 'in my head' version. So, for all we know — and especially in reference to anything even remotely suggesting participation in a crime — none of it is even true anyway.

I am grateful for the opportunity to leverage *The Sapient Salesman* as an outlet to share with you my inanity. As you sample the novelty, the wisdom, and [hopefully] the humor of these tales — please — enjoy the *Schadenfreude*.

STORIES ARE *NOT* PRESENTED IN CHRONOLOGICAL ORDER.

The Sapient Salesman

Spinning Life into Lessons, One Tale at a Time

To Forgive or Permit

When I was a little girl I used to sneak spoonfuls of ice cream when my parents weren't watching. I got away with this scheme for years because I never left evidence of my crime, but I fear the antics threw my spoon karma out of balance. You see, my parents ran the freezer pretty cold which resulted in rock-hard ice cream, the kind that spoons were never meant to dish directly. So rather than trying to concoct a steady stream of stories to explain why the sink was full of spoons that looked like they moonlighted as props in *The Matrix*, I just threw them out.

I remember the day, maybe seven or eight years ago, when my dad — in response to a comment regarding the surplus of spoons now occupying his kitchen — said:

> "MAN, I DON'T KNOW WHAT HAPPENED... FOR A WHILE THERE WE SEEMED TO GO THROUGH SPOONS PRETTY QUICKLY, AND RIGHT AFTER I CAVED AND BOUGHT A BUNCH, THEY STOPPED DISAPPEARING."

While I did confess — explaining how once I got strong enough to bend the spoons back, I didn't have to throw them away anymore — the universe still saw fit to give me a taste of my own medicine. Within weeks of the Edward Don outlet closing its doors in Chicago, my personal spoon supply suspiciously began to dwindle.

I later came to find out that Calvin, my husband at the time, employed similar argument-avoiding logic. You see, Calvin is afraid of the garbage disposal. Irrationally afraid. According to him there are hand hungry kitchen trolls hiding in the cabinets just waiting for the moment when he'd reach into the garbage disposal to check for rogue utensils to engage their remote control activation device. No matter how many times I assured him that the 'blades' in the Insinkerator weren't cut-your-finger-off sharp, Calvin refused to check before activating the machine.

Without fail, about once a week, I'd lose a spoon to his paranoia. If you're not familiar, garbage disposals do to spoons what cheese graters do to sponges; the result isn't pretty. So, rather than face the wrath of heckles from his wife, Calvin just started throwing away the chewed up spoons before I could assess the damage and mock his wussiness.

Now, faced with a shortage of one particular spoon model — and no retail source to supplement supplies — I've resorted to restaurant recon. Finding my pattern has proven difficult. So you can imagine my delight when, during a recent jaunt to Orlando, dessert got served with my spoons!

Before any of my meal-mates could even dive in, I declared my intention to pillage the flatware, and instructed everyone to lick their spoon clean when they finished. Frank, the least fun of this misfit trio, initially thought I was kidding. Herber new better. Herber and I had been working together for a while at this point; this was neither our first meal, nor the most ridiculous request he'd heard from me. Herber was completely on board.

> "I'M SERIOUS!" I SAID WHILE STARING AT FRANK, "I'M GONNA NEED YOUR SPOON. DO NOT GIVE IT BACK WITH THE PLATE!"

An argument of sorts ensued. Well, more of a debate really, regarding whether or not the restaurant staff would notice the missing spoons. Herber and I assured Frank that, despite my lack of a cloaking vessel – a.k.a. a purse – no one would notice three spoons in my back pocket. And if they did, worst case scenario, we could just run for it. We couldn't imagine that the Orlando PD would respond to a 911 call about three traveling salesman making a break for it with dessert spoons with any real urgency.

Frank never really got fully on board, but agreed that if we let him leave first, he'd pretend he didn't know what was going on behind him. So, finally, I casually giggled my way out of the restaurant with three spoons subtly tucked away in my jeans. As we made our way to the car, I attempted to share the backstory with the crew. But by this point the dessert's sugar kicked in and rocketed my giggle fit into high gear, so I'm pretty sure that through the laughing, I was the only one who had the faintest idea what I was going on about.

My overwhelming joy must have inspired Herber though, because before starting the car he turned to me and asked if I wanted more spoons. Of course I did! And before Frank could even renew his desire to *not* get arrested, Herber vanished back into the restaurant.

Moments later he emerged with the five additional spoons I had requested – I figured eight was a nice set-like number of spoons to leave with. I assumed, given their neat little napkin package, that they were snagged out of a bin he serendipitously cased on the way out. But I was wrong. We soon learned that Herber took a far less sketchy approach – he simply marched up to our waiter and said:

"I'M GONNA NEED FIVE MORE SPOONS!"

Dazzled by the directness of the request, the waiter was happy to oblige.

And here we are. Faced with the age old question: is it better to seek forgiveness or ask permission? I now have evidence suggesting that both strategies have merit, but even though my method did lose with a spoon-score of five to three, I'm not ready to switch camps just yet.

So next time, play to your strengths. My confidence that a long, laughter fueled, explanation would get me out of any potential trouble, has led me to a life of successful spoon smuggling; while Herber's confidence in his closing skills took him down a more noble path. You know what you're good at, so just do what you do. I'm sure you'll end up well spooned too.

Underwear Models

So I'm sitting on a flight from MIA[1] to LAX[2] next to a lovely gentleman whose first words to me were:

> "DON'T MIND THE GLITTER, I'M HERE WITH MY TEAM OF ANDREW CHRISTIAN UNDERWEAR MODELS AND WE JUST FINISHED UP AN EVENT."

I knew exactly which event he meant; I actually had stopped by the festivities on my way home the day before. We chatted for a little while, but before long he drifted off to sleep, leaving me alone with nothing but my thoughts for five whole hours.

Having exhausted my instant message quota, caught up on email, and tweeted my heart out, I dug deep into the Internet – searching for something novel to occupy my mind. But in the end, I decided to revisit an old friend: ListenToAMovie.com, who – much to my delight – recently expanded their catalog to include stand up comedy.

I turned on some Robin Williams and poured myself back into work for another half an hour before 'restless Erin syndrome' kicked in, hard. In a flash I realized I hadn't stretched yet, so I leaped to my feet with such shift enthusiasm you would have thought something just bit me on the bum. Mind you I'm in the front row of coach; I had the whole plane as an audience.

With the Chromebook perched on its side so as to maintain

---

a successful headphone-cable distance, I was ready to start stretching. About ten minutes into this limbering, a Robin Williams induced afternoon giggle fit ensued. Before long, I had tears running down my face and was bent over more from the laughter than the faux yoga.

This is about when I realized the other passengers — who naturally can't hear the jokes and who probably don't find the news segment airing over my shoulder particularly amusing — must think I'm insane. As I watched two women four rows back laugh with (or perhaps at) me, I couldn't help wonder how often salesmen find themselves in similar situations: standing in front of the room broadcasting our agenda on a channel our customers aren't tuned to.

Do we struggle to sell because our products are complex? Or because we speak in an alphabet soup of initialisms that render prospects too perplexed to purchase? How can we accommodate naivete without spiraling into a condescending consultation?

Deep down customers long for educational encouragement, they just don't want to admit it. In fact my most successful sales engagements are those when I inspire the customer to ponder their circumstance. This shouldn't come as a surprise. Just think about the plethora of prospects who flock to lunch-n-learns, trade shows, TED talks, and Twitter to source solutions from today's thought leaders.

So next time, try to teach your customer something. When you stop selling for a second and share some insight about their business, you'll leave them wanting more. Before you know it, you — like Mr. Williams — will find customers lining up to hear what you have to say next.

Window Shopping

When I relocated my desk from the living room to the den in my house back in Chicago, I was reminded daily what a difference new windows make. Which, in turn, reminded me of the shenanigans that ensued the year before while window shopping. I entered the purchasing process the same way most of our prospects do: relatively naively. Much like our customers, I armed myself with some misconceptions and went out in search of quotes. Three vendors volunteered to come evaluate my window needs.

The first guy walked in, surveyed the space, and asked where my − now ex − husband was. I explained that, as it was 2pm on a Tuesday, Calvin was at work. This caused the guy to stonewall me. Before even asking who the decision maker was, or discussing what we were looking for, he deemed a conversation with me insufficient and suggested he come back another time.

The second gentlemen had similar qualms about selling to just me, but pressed on. He came with binders! These binders housed material detailing how their windows knocked the socks off the competition. We painstakingly reviewed the presentation, page by page, until he believed that I was sold on his brand. When we finally arrived at the portion of the conversation I cared about − how to address the window design − he turned to me and declared that the space was too unique for him to opine on and he needed a second meeting; he wanted to consult with his design team.

Later that evening window guy number one returned and

this time, he brought props. He began by trying to teach me how windows work. So I quickly dusted off my cursory understanding of thermodynamics and started asking questions. Unfortunately for him, his science-light sales approach consisted of telling me how one gaseous filler was 'better' than another — differentiating both from within his own product line and the competition — but he had nothing in his bucket of canned responses to back up his claims.

I asked him two or three relatively simple science-type questions about the thermal conductivity of the gases he mentioned and he panicked. Panicked! Mind you, this whole time he directed the pitch at Calvin who sat there silently trying not to laugh. But just like contestant number two, he failed to discuss the problem we were actually trying to solve. He had no opinion on design.

The next day salesman number three — a tall man named Iver — arrived. His opening line was a question. Well, a compliment and a question.

"THIS IS A GREAT SPACE, WHAT WOULD YOU LIKE TO DO WITH IT?"

Music to my ears!

It only took four meetings and three companies for anyone to ask why I called. He listened, provided Calvin and me with several off-the-cuff options, took measurements, and asked that he be given the opportunity to review the specs with his design team before making a final recommendation and discussing pricing.

The subsequent meetings with Iver included conversations about the science and comparable benefits of various pane designs. He carefully weaved a differentiation story into the presentation of the final design and ultimately gave us the context we needed to move forward with our decision.

Calvin and I did ultimately obtain pricing for Iver's design from the other two companies. Even though Iver's bid wasn't the lowest and his company was the smallest of the three, we went with the only firm that seemed to have any clue about what we wanted out of the purchase. And I couldn't have been happier with the result.

So next time, be cognizant of your place in the sales cycle. Your differentiation efforts will fall on deaf ears if the customer doesn't first believe: that you understand their needs, that your solution can solve their problems, and that you have the facts to back up your claims. Remember – for a smooth trip around the cycle – listen, pitch, differentiate, close.

Joint Custody Street Frogs

So I'm walking down Washington Ave. in South Beach the other day with Ralph — my best mate — when all of a sudden I notice he's no longer by my side. This is unusual because he's one of the few blokes I walk about with who can actually keep up with me. Plus Washington isn't that interesting of a street, scenically speaking — it's mostly boring beach stores and dingy bars that won't be open for hours.

When I turned around to see where I lost him, I discovered him standing by the curb, staring at a puddle in the street. Ralph's a weird kid, don't get me wrong — that's why we get along — but this posture was weird, even for him. He looks up flashes me his signature focused-interest look and goes:

"COME CHECK OUT THESE TADPOLES!"

Leave it to eagle eyes to spot a school of tadpoles swimming around in the gutter out of his peripheral vision. We both immediately became enamored by the idea of street frogs and quickly boarded the Snowball of Bad Ideas and Giggles, or 'SoBIG' as I prefer to think of it.

"WE SHOULD TOTALLY RAISE THEM!" I SQUEALED.

He didn't immediately say no. I found that fact very promising since we both knew they would have to live at his place; I travel way too much to grow frogs. We eventually carried on our way, but as the afternoon marched on and our respective errands concluded, I couldn't get the street frogs out of my head.

After a few more rides on the SoBIG the excitement surrounding this adventure became *so big* we simply had to go back. Armed with two gallon zip-top bags, a couple of plastic measuring cups, a strainer, and a smile, we marched our way back up Washington to 'rescue' some tadpoles. We might have gotten carried away. When we returned to my place and transferred our refugees into their five gallon bucket halfway house, we realized we retrieved almost two dozen taddys.

Sadly I had to leave town the next day and head to a software/sales boot camp for a new gig. But like any good absentee parent would, I first established a solid joint-custody arrangement; Ralph committed to their care for the next week while I was out of town.

Flash forward a couple days. It's barely 8am and I'm sitting, under-rested, in the training where the instructor asked everyone to include "something personal" as a part of the go-around-the-room introductions. The ten some-odd people preceeding me spouted off parallel stories about their two and a half kids, white picket fences, and wholesome households. So not much correlation or inspiration there. Plus it still being the crack-of-dawn and all, my brain was far from fully booted.

I was stumped! What possible story could I tell? Surely something funny happened recently that would sufficiently compensate for my family-light, Miami-like lifestyle. I replayed my whole weekend and the only two stories I could think of were: one about throwing a rock at a coveted coconut, and the one about the tadpoles.

As I listened to the man to my right's introduction which included a declaration of his victory over cancer, I realized I was in a no-win position. I mean how do you follow cancer guy?!?

The coconut story was clearly out. Not only is it way too aggressive, it doesn't end with me putting a lime in a coconut. Salespeople like to win; I needed a victorious story! So I figured screw it, the frog story is funny and perhaps the taddys would lighten the mood some, so I said:

> "HI I'M ERIN, BASED IN MIAMI BEACH, HAPPILY DIVORCED AND ...
> SO ... THE OTHER DAY I WAS WALKING DOWN THE STREET WITH A
> BUDDY OF MINE ..."

You know the rest. It played well (or so I thought); the room laughed at the punch lines, asked to see the pictures, and really seemed to enjoy the brief tale. Cool, right?

Yea ... not so much.

I come to find out that this became the talk of the dinner

outing — which I had skipped on account of my severe lack of sleep the night before. Evidently instead of finding the story charming, with a touch of do-gooder, everyone now considered me bat-shit crazy. I would further come to be known as the "crazy tadpole lady."

Fan-tastic.

Look, I'm a quirky bitch, but I'm not actually crazy. The way I figure it, in life, sometimes you just do it for the story. That said, however, this also isn't the first time I've had a story backfire on me at work this spectacularly, and I couldn't help but wonder if I'd ever get good at reading the room. One might argue that the frog story was a little much for a first impression, but all things considered, I still find it pretty benign. However, I am willing to concede that — despite my disagreement with the consensus — there's probably a lesson here.

So next time, play to your audience. In the face of folks fitting the 'formula', stick to stories of simplicity and stodge. When you reserve your trenchant tales for clearly kindred spirits, you just might receive the warm welcome you wanted.

Nice and Happy

On my flight to Dallas the other day, I got to talking with a dead heading captain about why I try to avoid connecting through the Metroplex[3]. We bonded over a few anecdotal examples of Texas' southern 'hospitality' and by the time we reached our cruising altitude, I had officially secured myself another single-serving friend. As the conversation carried on, I decided to share an epiphany I had during a recent trip to Salt Lake City about 'nice' people.

You see, generally speaking, I'm a happy person. I smile, giggle, and even occasionally jump up and down; I get — and often am — excited about life. Yet even as a turbo extrovert, who longs for conversational companionship, I can — as it turns out — be happy ALL BY MYSELF.

Nice, on the other hand, requires others to participate. You can't just go around being nice to yourself. Nope. If you want to 'be nice' you need to involve someone else. You require a target for your niceness. What bugs me the most though, is the impact on the unsuspecting third party, especially when that party turns out to be me.

Think about it, if while walking down the street, minding your own business, you encounter a happy person in whom's happiness you choose not to participate, nothing changes. They go on being happy, you go on walking.

However, if on that same walk you cross paths with a 'nice' person who decides to direct their niceness at you, and you

---

[3] A NICKNAME FOR DALLAS FORT WORTH INTERNATIONAL AIRPORT (DFW)

choose not to participate... they get frustrated, and – somehow – you end up walking away an asshole.

This emotional codependency got me thinking about the definition of 'value add' applications. Can an application that requires a client to purchase something else first, really be considered valuable? Is it enough to solve problems in a silo, or does sustainability require sharing and technological companionship? If a single philosophy prevails, should we switch to exclusively selling software of that sort?

Just like happy isn't intrinsically better than nice, I don't think standalone blanketly beats plugins. Clearly both business models can, and do, thrive – it just depends on how they are applied. Similarly to how happy people are more receptive to the advances of the nice, clients already subscribing to the software are better suited for a plug-in pitch. And apathetic people, like empty handed prospects, shouldn't feel obligated to go out of their way, to buy more than they set out to, just to justify your business model.

So next time, sell selectively. For any solution to be classified as such it must allow a customer to derive direct value. When you focus on prospects participating voluntarily, instead of forcing future features, not only will you close strong, you might just make your clients happy, and that's ...

... nice.

Advertised Price

You know how sometimes you start cleaning your house — just an ordinary tidying, putting away shoes and whatnot — when all of a sudden you find yourself elbow deep in an air conditioning vent holding a softball-sized block of plaster?

Just me? Okay.

Neverless this is precisely how I found myself one summer afternoon — sitting on the floor with my hand stuck inside an air return vent, clutching a clump of plaster. Upon further investigation, many of my cold air returns were similarly debris ridden. So I went into the following endeavour knowing full well my vents could stand a brush cleaning, because I doubted a vacuum alone would provide the power necessary to the remove large rocks that lay beneath.

On a separate note, working from home has left me with some old-lady habits. For instance, I now regularly thumb through the weekly junk mailer coupons. I'm amused by how the price of the same product fluctuates so wildly from week to week; I find myself eager to see what bullshit gimmick these companies will try next.

Over the course of a month, I noticed a steep downward price trend in air duct vacuuming. The price for a "whole house duct cleaning" declined from $300 to $150 in a single week. So a few weeks later, when a company's coupon hit twenty-five bucks, I pulled the trigger and called for a cleaning.

The coupon clearly stated the service only included vacuuming. It instructed us to call for pricing regarding any additional services, like sweeping or sanitizing. Clearly this coupon was a part of a lead generation scam; I prepared myself accordingly.

On the day of the service, I opened the door to discover a 5'3", 120lb, blonde, white guy. He came in and quickly surveyed my duct work while his comrade set up the vacuum. He scoped the vent in the living room and – shockingly – unveiled some rocks. I acted surprised. Mistaking my smile as an indicator of complacency, he immediately began to explain all the products they offer.

After every couple of products, he'd pause and pitch me on the health of my house – or lack there of, as it were. Halfway through his catalog of services he would have me believe I was lucky to be alive, sitting in there all day, breathing that dusty air. I tolerated this charade, and not just because I knew it had *Sapient* material all over it. I actually wanted all the vents brushed – not just vacuumed as included with the coupon – and I enjoyed how he kept volunteering all his would-be counter arguments in advance of my objections.

The negotiation started at $750. My disgust prompted an immediate concession; since I called "off the coupon" he could do it for $375. I said no. Then he explained – again – how, since I work from home, I should really care about the quality of the air I breathe. I agreed my vents were dirty, but explained I was only interested in getting the base vacuuming at this time. He countered with $280.

Calvin and I decided $250 was max we were willing to pay

before blondy boy even arrived, but who doesn't like to play a little 'limit access to the fictitious decision maker?' So I told him I needed to consult with the hubby.

This lie gave me an excuse to go back to work. Within five minutes, a broom on a drill walked into my office and, as I spun my chair about, a miniature man appeared from behind the tool; he sought complacent consent and asked if he should just start in there. This kind of pissed me off. I explained that Calvin said no, and we were only interested in the basic service and a quote at this time.

My response didn't go over well.

Vent guy returned and started lecturing me. He asserted that I called because I wanted clean duct work. I explained that my expectations weren't that high; I would gladly settle for clean-er vents. He then tried to make me feel bad; stating how $25 wouldn't even cover his gas to get here, and I suggested that might have been poor planning on the part of the marketing department, but ultimately their fuel expenditures were not my problem. I could tell his frustration was turning to fury so I threw him a bone.

ME: "LOOK I'M SORRY WE JUST DON'T HAVE BUDGET AT THIS TIME."

HIM: "SO IT'S ABOUT PRICE!"

ME: "MORE ABOUT TIMING, I ASSURE YOU."

I promised to use his service at the end of the summer. He offered $200. Again, I declined. I grew so bored with his desperation that at this point I actually started selling him some software. I suggested that if he had a proper CRM[4] he could just enter notes right here instead of running off to tell

---

[4] CUSTOMER RELATIONSHIP MANAGEMENT (SYSTEM)

the corporate office how it was going every five minutes. He bit, albeit probably just to placate me, so I handed him a card and walked away saying I would try Calvin again.

After regrouping he approached one final time. I told him we choose to pass, and he said he wasn't going to take no for an answer.

"THAT'S UNFORTUNATE!" I DECLARED "NO IS THE ONLY ANSWER I HAVE, AND I'M REALLY GOOD AT GIVING IT."

I further declined his $150 offer, and he was all "look I don't want to have to come back, it costs ... blah blah blah..."

I assured him that it was okay with me if he didn't want the further business, but should he change his mind I would take his business card and call him at the end of summer. He then asked if I would do it for $100. After a long pause I tried to say "fine," but he interrupted to declare they could only go as low as $150.

At last the bottom! I made him wrap the $25 coupon into it, throw in sanitizer — which I didn't even want, but I was having too much fun — for an out-the-door, total price of $150; take it or leave it. After some balking, he took it.

All of my vents are now shiny and clean, and I got to spend the day walking a sales guy into every trap in the book. Not bad for a Monday afternoon.

So next time, beware of customers like me. Prospects who — despite having a genuine, confirmed need — systematically revoke every negotiation point you have will drag you into a price war. Remember, sometimes it's better to walk away than to waste your time chewing away at your margin.

Pesky Pigeons

I'm having pigeon problems. Not Pidgin problems, I'm talking actual birds here. What's worse, there seems to be a consensus among my friends that I shouldn't just shoot them. So I'm stuck resorting to more benign tactics. The thing is though, I don't speak pigeon and — after the last week — I'm of the opinion that pigeons don't respond to reason, in any language.

Here's how it went down ...

Google provided me with a list of things pigeons don't like. I reviewed it, considered what my many years as a bird owner has taught me about bird preferences, and decided to open with a 'sticky feet' attack. So off to the Walgreens I went to purchase a giant tub of knock-off Vaseline and a can of WD-40. (I realize WD-40 isn't sticky, but evidently the smell bugs them quite a bit.) I returned home, gloved up, and lubricated the railing on my balcony.

While I basked in the glow of my cleverness — giggling and rubbing my hands together — and waited for my first avian victim to

land, I realized I was more likely to catch a friend's forearm than a bird. But, alas, these are the risks we have to take.

As the night waned on, and the sun went down, I forgot all about the Vaseline. The next morning I woke up to cooing. COOING! Confused, I worked the foreign sound into my dream for probably a half an hour before rising to find a pissed off pigeon doing the gooey foot dance on my balcony. I chased him away and after forty-eight hours of no further pigeons, I declared victory.

I spoke too soon. The pigeons took the time to contemplate alternate landing paths. This time they approached in such a way as to bypass the top railing altogether, roosting directly on the table next to my SkyMall meerkat statue.

::sigh::

Time for more Vaseline! So I gloved up again and bolted outside. Meanwhile, Shale — a beach-cop buddy of mine — is sitting at the counter working on his laptop. Little did I know the pigeon I shooed away wasn't alone. Just as I began to smear on the next layer of goo, another beast of a bird comes flying out of from the milk crates under my statue. I squealed like a bitch and startled the bird, which then flew inside my house.

There was a pigeon INSIDE MY HOUSE!

And Shale? Shale doesn't do shit! I, still squealing, bolt back inside to try to box it out and prevent any aviary advancing. It hit the window pretty hard — an event I briefly cheered until realizing it didn't die — before finding its way out.

Oh, now it was really on! We're way past WD-40 at this point. I exhausted the tub o'petroleum jelly, topped it off with a large bag of chili powder — which, rumor has it, birds like even less than penetrating lubricants — and chastised Shale for not drawing down on the intruder.

The next day I, again, awoke to cooing, a sound I now believe to be a pigeon's pity plea. But thankfully, since I shooed that bird away, none have returned. If they do come back though, I'm seriously going to shoot them — first with rubber bands, then pellets, then nets made of rubber bands covered with WD-40.

With my contingency plans in my pocket, I started to think about how unreasonable the birds were being by not taking the incredibly clear hints I had smeared out for them and how often we find ourselves in a similar situation.

As Sales Engineers, we speak a different language than our prospects. Very few clients carry a geek-to-human pocket dictionary. Instead they sit quietly — probably cooing on the inside — and wait for us to leave before seeking a solution that's easier to swallow. So short of tranqing the prospect and spraying them with enough backstory for them to finally understand what you're selling, how can we effectively communicate with all those liberal arts majors out there?

Simple, crank up the sales and speak their language.

So next time, find common ground. Take an inverse-pigeon approach: keep it simple, figure out what your prospect likes, and anchor to it. When you can tie your actions, your product's features, and your company's value to their desires, your prospects will certainly come home to roost.

Don't you know who I am?

Failing to unify a company on a single, thoughtful, customer relationship management system doesn't just introduce inefficiencies — it can make your organization straight up irritating. Don't believe me? Consider this vehicular tale...

I'm a loyal patron of Grossinger Toyota North in Lincolnwood, Illinois. Not just because fifteen years ago my mother wrote a sternly worded letter to Toyota corporate granting us a lifetime 10% off all service: parts *and* labor. Because, perpetual coupons aside, Grossinger does good work; my family happily patronizes the dealership. They even sold me my first [new] car back in the day — a 2002, blue Rav4 with only three miles on it.

So when I found out I'd be moving to California and decided to retire my Rav, my first instinct — naturally — was to stop by good old Grossinger and solicit an appraisal. After waiting ten minutes for the sales guy's computer to boot, he punched my information into some antiquated application. From his flickery CRT monitor we managed to uncover the going rate for — what basically amounted to — a ten year, 67,000 mile rental.

After spending the next week out of town, I returned, title in hand, to close this deal. While waiting in their sales pit, my phone rang. Who could possibly be calling me right now? But a friendly Grossinger lead qualifying lady, of course! I accused her of stalking by asking if she could see me sitting in the lobby. She didn't get the joke — even after I explained it. But that didn't stop me from suggesting she should use the remaining call time to help me jump the queue. The call ended after her confusion exceeded her patience.

Flashing forward – I sign the papers, take a quick business trip, and return ready to fetch my check the following morning. Tucked away in my pile of missed mail, I find a letter from Grossinger expressing their interest in purchasing my 2002 Rav4. And for $800 more than the negotiated price, no less. Knowing full well my ride was far from 'mint condition' I excused that detail in favor of inspecting the postmark; sure enough, the letter was sent five days AFTER they took title.

The charm continues.

Over the next week, I get three emails and two voicemails from agents wondering if I was interested in discussing a trade-in, or perhaps even to "get me into a new Corolla." What raises the ridiculous factor even higher though, is the fact that during the process, I explicitly asked to opt out of all email communiques and the car I suggested they put in the 'mandatory' next-vehicle field was *not* a Corolla.

Then one afternoon, a few weeks later, I got a letter reminding me that the Rav is due for its 100k mile service. My head nearly exploded. Not only do they own the car now, but holy hell, if anyone should be aware of the incredibly low mileage of my ex-vehicle, it's the people who serviced it for the last decade!

So next time, remember everyone needs CRM. Everyone. I realize simply having a system doesn't guarantee you will use it, let alone well, but customers expect more these days. Use your sales tools; be a better salesman.

A Picky Packer

So I'm packing. Again.

This makes the sixth time in five years — five times moving my crap, once removing Calvin's — and each time I find myself trapped in a guilt loop: think giggle loop but far, far less fun. What I really want to do is pack up all my clothes and just ditch the rest. The simplicity of the binary condition appeals to me; it would be so nice to say:

"IF I CAN'T WEAR IT, IT STAYS IN CALIFORNIA."

But just as these thoughts cross my mind, the lesser used portions of my brain unexpectedly activate in one of three irrational forms: guilt, frugality, or laziness.

So I start cherry picking items that are 'worth shipping,' and before long I'm buried under a snowball of shame. There I am recalling, unsolicitedly, every detail of the circumstances surrounding the acquisition of my possessions. God forbid should someone walk in and inquire as to the where, when, with whom, for how much, and why behind my purchase decisions because that... that activates some supremely girly guilt.

Every. Single. Time.

The thing is, I don't feel bad about leaving things behind; in most cases I had forgotten I even owned the items in the first place. What tickles my human bone is the fact that I feel like a sociopath because I can recall all the details but I don't actually feel any kind of emotion in response to the internalized story. What's worse, it all happens so quickly that onlookers mistake the emotive outburst as sentiment for the possession and not narcissistic disappointment and shame: the staple fuel of the guilt loop.

Great, so Erin's a crazy person, no surprise there...

But what about the every-man: the poor schmuck you're trying to convince to participate in your product's alleged benefit? If you sell a product born in the era of hyper-elastic expectations, then you probably bombard your prospects with details too. You may not force them to weigh the value of their stand mixer against the cost to ship it, but you probably subject them to a plethora of seemingly arbitrary decisions.

Remember the first time you encountered a Starbucks? That moment when you realized you were about to have to make a billion decisions about size, caffeine, dairy, and foam? Well, sometimes customers just want a cup of coffee – not a line, a vocabulary lesson, a hundred options, and the judging stare of a pompous barista.

So next time, simplify the sale. Remember, more often than not additional options or details only give your prospects more reasons to second guess their purchase decision. When you stop providing your prospect with potential objections, you might just find the time to go for the close.

Dialing for Doctors

I decided — in the spirit of committing to my new zip code — it was time to secure local healthcare. Because as much as I enjoy the excuse to fly to Chicago whenever I need my teeth cleaned, it's not terribly practical. After looking online, cyberstalking various providers and scouting the potential commute, I reached the point where I was going to have to talk to an actual live person.

I hate making these kind of cold calls. The number is usually shared between a group of doctors, which forces me to butcher someone's name before being connected. Plus, once I am, it's 50/50 whether I'll actually get to complete the task of making the appointment on the first call. So, I did what any nerd facing such a task would do — I procrastinated until I could think of a better plan.

Unfortunately I didn't come up with anything. So I eventually bit the bullet and dialed for doctors. Five offices, four days, three blind disconnections, two unreturned messages, one validated point, and zero appointments later... here we are.

Pouting, I turned my attention to Twitter, where I stumbled on an article about how customers admitted their willingness to vendor hop when presented with a vendor alternative that provided more, or more preferred, communication channels. "Oh yea, ZocDoc!" I thought as I opened another browser window, logged in, and updated my insurance info. My search returned a single doctor, on the beach, in my insurance plan.

Bam!

She was attractive, had no disparaging reviews, went to a school I'd heard of, donned a last name I could pronounce, and had availability on Monday. Perfect. I clicked away and locked down a time. No receptionists; no voicemails; no worries.

I realize my doctor selection criteria is unorthodox, but the fact is you probably have gone head-to-head with prospects with equally odd yardsticks. So quirkiness aside, ask yourself − what barriers to buying do you throw up? How many hoops do you force your customers to jump through just to place an order? Do your clients invest the bulk of their time in the right parts of the prospecting process?

As a person who personifies the modern customer whose choices are driven by channel and convenience, you may be inclined to dismiss my warning. Do so carefully, because even analysts agree these numbers − the percentage of customers like me − will only increase over time, especially as more and more people realize their preferential stubbornness will be placated, if not outright rewarded.

So next time, stop blocking the money. Take a moment to ask your customers, and the ones who went another way, how they wanted to buy. When you offer solutions via the channel people want, you remove the handicap and will close more, and more loyal, business.

Determined to Donate

My night went something like this: realize I've been in Miami six months, vow to reconsider the cupboards of shit in my apartment that have gone unaccessed, start tearing through the kitchen. The easiest project to tackle was the booze and book cabinet, so I did. Don't worry, no liquor was lost during the execution of this mission.

After review, my once mini library turned micro and twenty-five pounds of overflow literature got loaded up for the library. I consulted Google to confirm location and operating hours of my destination, strapped on a backpack, snagged my skates, and took off.

The rollerblades lasted about five blocks. In just under a half a mile I became convinced the displaced center of mass would land me on my ass. To stubborn to turn back, I plodded the next fifteen blocks in flip-flops... in vain. Upon arrival, through the sweat dripping off my brow, I saw a sign. A sign saying the library closed at 6pm.

I tried to play it off – managing to maintain stride as breezed past the door and turned the corner. Hiding along the side of the building, I put the bookbag down, drank some water, and regrouped. The denial continued for a good thirty seconds during which time I reviewed the entrance three or four more times; I couldn't believe there wasn't so much as a book drop slot for me to unload – albeit anonymously – my donation into.

Schlepping these books back home was so *not* an option!

"Now, should I leave the books on the stoop, or convince a stranger to take care of it for me?" I thought.

Since it rains a lot in Miami and books are — you know — made of paper, I decided to give option number two a try first. I quickly identified three targets: the Miami City Ballet School, the Bass Museum of Art, and the two random cops parked on the other side of the park. The ballet school was closest so I skipped over, ascended the stairs, marched up to the security desk and said:

"Hi! This is probably the dumbest request I've ever made of a total stranger, but this is the thing…"

After a ninety second recap of the events leading up to this moment, the guy could barely keep from laughing. I almost had him convinced it was perfectly reasonable to keep a bag of books under the desk overnight with the promise of a sweaty lady on rollerblades returning during library operating hours. But under the peer pressure from the two blokes standing off to the side, staring intently, and waiting to see how this played out, he hesitated. In an effort to keep him on the hook, I carried on:

"Look I don't really care if I get the tax receipt, you can donate them for yourself for all I care, I just don't want to carry these all the way back to 5th Street!"

Breakthrough! The dude directed me to the heavily disguised drop bins hiding in the parking lot behind the building. Elated, I sauntered over and donated away. As the books banged down the box two by two, I couldn't help but curse Google a little bit. This isn't the first time they've done this to me — suggested an establishment would be open when, in fact, it wasn't. Before they added their little side box of 'information,' I'd click, consult the company's site, and presumably avoid this situation. I wondered if similar conditioning also causes our clients to believe the *first* source over the *best* source.

In sales, we talk a lot about getting into deals early, setting the groundwork, making the plan, positioning against the competition, and locking into 'column A,' but does that matter? Do our prospects 'trust, but verify'? Or does the word of the first source permanently taint their taste for a solution?

Eventually I'll learn to stop assuming Google knows all and go back to calling before rolling on out, but for me, the investment is only of time and travail. For our potential customers, the cost of ill advised commodities impacts both contracts and careers.

So next time, confirm they believe you. Remain a credible, honorable, and up-to-date information source for your customers. Because remember, when you become your prospects' trusted advisor, you control the deal, you control the plan, and you control the close.

Who's Selling Whom?

Recently I had the opportunity to witness, first hand, a formal battle between 'flirting for sport' and 'flirting with purpose.' But first, a little backstory…

I've recently come to befriend a group of vice detectives that I affectionately refer to as the Coterie. They work primarily in plain clothes and let me tag along from time to time. As someone who moved to Miami literally knowing no one, this is a nice arrangement for me; I get to meet a lot of great, random people while enjoying countless cocktails and convivial company, and they get to look less suspicious by having a girl in the group.

Win, win.

So back to Friday, the boys were out hunting for hookers. They took turns perp-walking the ladies back to a hotel room they had set up for making arrests. This week in particular was more action-packed than normal because resources from sister agencies and departments came out to lend a hand. Expectations were high and everyone was eager to find themselves a fish.

While Shale, the leader of the Coterie, did his thing across the bar, I took the opportunity to catch up on some television. Leaning on a table, I toggled my focus between Shale's pickup efforts and Sports Center. Moments later a very tall bloke shimmied past me. He mumbled something about my shoes as he made his way to the bar. Never one to forgo an opportunity for extroversion, I requested he repeat himself which initiated a conversation of sorts.

You should know, at this point I'm like 90% sure he was at least tangentially affiliated with the Coterie. But even if he wasn't, the fact that I was literally surrounded by cops left me feeling safe and – dare I say – a little bit cocky. So when he turned conversation from friends and footwear to one of sex and solicitation, I couldn't resist playing along. I figured either he was with us and this would be hilarious, or he wasn't, and I just might prove to be more useful than your average cover-chick.

Contrary to popular belief, you don't actually need to bang the ho to arrest her, but there are certain elements of the crime you must establish. Just like in any sales situation both parties must agree on price and product – or in this case, service. Normally at least one party is actually trying to transact, but in this exchange I had nothing to sell and he had no intention to actually pay me.

I later came to learn that this was Detective Jim's first time and he had no idea who I was. The conversation that ensued pegged sport-flirting against work-flirting. It was brilliant.

I knew better than to agree to anything; he knew that I should be the one to say everything; neither wanted to cave. He started by laying out some story about being in from out of town, but wasted no time asking me up to his room.

"WHY SHOULD WE GO UP THERE?" I ASKED NAIVELY.

"TO PARTY," HE REPLIED SUGGESTIVELY.

I rolled my eyes and didn't reply for apparently a long time because before I knew it, he broke the silence.

HIM: "COME ON, WE CAN GO TALK."

ME: "WE ARE TALKING, WHY CAN'T WE JUST TALK HERE?"

I smiled at how cheesy this was and said:

"SO WHAT DO YOU DO?"

He delivered some story about how he was a professional basketball player from Europe. Jim's entire plan must have hinged on him being tall because he couldn't name a single team in the league out there. I say couldn't, but he alleged it was more a wouldn't situation since "[I] probably wouldn't have heard of them anyway." This bored me. So when he tried to divert the conversation to me, I let it happen.

I vowed in the beginning, for the sake of truth in sales, to answer all his questions honestly. Mostly to see if I could still social engineer this conversation to a close without having to lie. But that didn't mean I wasn't prepared to sprinkle in a little spin along the way.

HIM: "SO WHAT DO YOU DO?"

ME: "I SELL SOFTWARE …" I SHRUGGED, SMILED, AND ADDED: "… IT'S A DAY JOB."

Mr. Happy Ears assumed I was being suggestive, not literal, and pressed on by returning the topic of conversation to his alleged hotel room. I insisted that I had no idea what he meant, and asked what − specifically − he wanted to do. This lead to our first of two Loony Toon-style conversation loops.

HIM: "TELL ME WHAT YOU WOULD DO."

ME: "TELL ME WHAT YOU MEAN."

HIM: "TELL ME WHAT YOU'D DO IF WE WENT UPSTAIRS."

ME: "THAT DEPENDS. WHAT DO YOU WANT TO HAVE HAPPEN?"

Round and round we went, all the while the smile on my face getting bigger and bigger. With each additional grin-wrinkle he became more convinced he was making progress. Finally he caved. He whispered a series of requests into my ear that would even make an experienced lady-of-the-night who spent her career servicing a ship full of sailors blush. I pulled away as he asked how much − mostly out of shock − to look to see if he was for real.

ME: "YEA, I HAVE NO IDEA WHAT THAT GOES FOR DOWN HERE."

This truthful declaration confused him, so I followed up with a quip about being new in town − also true, for the record. This threw us into our second, and final, episode of Looney Toons.

HIM: "SO HOW MUCH?"

ME: "SERIOUSLY, I HAVE NO IDEA WHAT THE GOING RATE FOR THAT IS DOWN HERE. WHAT ARE YOU OFFERING?"

HIM: "I'M NOT GOING TO OFFER, YOU HAVE TO TELL ME HOW MUCH."

ME: "HAVE TO?"

HIM: "YES, I CAN'T JUST PICK A NUMBER."

ME: "WELL THAT'S ROUGH, 'CAUSE I CAN'T BEGIN NEGOTIATIONS WITHOUT AN OPENING OFFER."

HIM: "SO YOU HAVE A NUMBER IN MIND?"

ME: "NOPE. DO YOU? ..."

This went on for a while. The best part is I never actually had to lie to him, even when I delivered the line:

"LISTEN BABY, IT'S FRIDAY NIGHT AND I HAD ABSOLUTELY NO INTENTION OF WORKING TONIGHT."

I could barely keep a straight face when he — so predictably — misinterpreted the remark. After about ten minutes my stubbornness prevailed. He finally threw out a number!

Flirting for sports wins again!

Just as I was recapping the deal, however, Detective Sixx — another member of the Coterie — walked up and sat down at our table. The poor bloke's eyes about jumped out of his head as he tried to shoo Sixx away. He really thought he was on the cusp of a close, but by this point I started laughing out loud; I couldn't play along any more. Sixx and I clued the dude in on the situation.

As we walked outside to join the rest of the team, I couldn't help but wonder how often we get so caught up in the pitch and the process that lose track of who's selling whom. Should we attempt regain control before carrying on? Or is co-selling with your prospect an opportunity for everyone to win? Can we sell successfully while we're being sold to?

I'm inclined to believe that while two alphas may enter a contest, only one will leave victoriously. If your customer does a better job of controlling the conversation, you're unlikely to complete your agenda. What's worse, the customer who is just playing − shopping for sport − has no reason to give you a go at the steering wheel.

So next time, look out for games. When your prospect is in it for the story rather than the solution, the deal − like your wheels − will spin and spin. Instead focus on customers with a genuine need, whose participation goes beyond that of sport. Contracts close when everyone plays with purpose, pursuing the same product at a true price.

Shall we bake?

So I've already bitched about the perils of packing, and that could have been the end of it, but it appears the universe was listening. To be clear, by 'the universe' I mean the mysterious and conniving underbelly of my subconscious whom, in the face of stress, always seems happy block and tackle. Luckily the resulting selective sensory input makes for the best stories. Case in point one Sunday...

I'm standing in the kitchen of my new Miami Beach apartment compiling a list of all the things I intend to pack and ship down. After only three days of zen-ful beachfront living the list was already well on its way to becoming quite abbreviated. As I mentally mapped my kitchen utensils to the new layout, I vowed to only allow for 10% utensil creep. That's to say, I'd only let myself bring one forgotten item for every nine that made the list. Basically I intended to compile a detailed and comprehensive list of 'required forks.'

Mind you, all the while I'm listening to a Law & Order marathon; then someone on the TV mentions baking. As I turn to survey the cabinet space again, I think: "Man... I didn't even consider cake accoutromount! That's going to be a long list!" It was in this moment, three days after signing the lease, that I realized there's no oven in the apartment.

I immediately started laughing out loud and called the only person who would find this as funny as I did: my mom. She puts the call on speaker so my dad can hear before asking: "how'd the hell you manage that?" It never occurred to me that someone wouldn't put in an oven, so I didn't even think to check. Every time I toured the place I was all:

"BIG CLOSET? CHECK. BALCONY? CHECK. HAPPINESS AND SUNSHINE? CHECK AND CHECK. SOLD!"

In retrospect, *House Hunters: International* did teach me that European households tend to be less oven-inclined and the presence of a bidet ought to have provided a clue, but oh well. The oversight provided much needed simplicity: any pan that's too big for my toaster oven stays in California.

As the laughter faded and the stress melted away, I wondered how our customers would handle a similar situation. In the face of a missing 'common' feature, would they embrace the opportunity to table superfluous use cases or flip out facing the unforeseen change? Is it in our best interest to point out potential shortcomings and manage expectations? Or should we assume the prospect's list of 'must haves' is comprehensive?

We'll never know how someone will react to surprises, but in my experience, most first-time customers are drama queens. At the point of sale the focus — and with it the risk — transfers from purchase to project, and your customer's reputation now hinges on your product making good on its promises — whether explicit, implied, or assumed. Be prepared to show them the silver lining hidden inside the inevitable surprises. I mean, I never bake anyway and now I have fewer boxes to pack.

So next time, remember most buyer's remorse happens while the ink is drying — when they once again have time to worry about the future. By supporting your prospect through this transition you'll accelerate the logistics of the close. Supported customers are happy clients.

Too Many Moving Parts

After my speaking session at Tech Day Asia — a three city traveling trade show — I found myself chatting with a couple of other vendors by the booth. All of a sudden very tall Asian man appeared. I remembered him because he was easily 6'4" and incredibly thin, weighing in at no more than 180lbs. Curiously, he attended my talk at the previous city too. Which begs the question: who travels from Singapore to Kuala Lumpur to hear the same content over again? As you might imagine, my weirdo alarms were firing on all flutes.

He lurked for a moment behind another gentleman then looked at me, looked at this skinny white guy next to me, and said:

"YOU TWO SHOULD GET MARRIED."

Mind you I had only just met this kid to whom he was referring; I didn't even know the white guy's name. So, clearly the tall bloke wasn't basing this assertion on any visible chemistry between me and the pale, bearded fellow; he simply believed that someone my age needed herself a husband, pronto. Much to my surprise, Jon white guy — as I later came to learn — played along perfectly.

ME: "WHO HIM? I DON'T EVEN KNOW HIM."

TALL DUDE: "YES, YOU TWO SHOULD GET MARRIED."

ME: "YEA, I DID THAT ONCE. NO THANKS, THAT'S A LOT OF WORK."

TALL DUDE: "YOU NEED TO BE MARRIED, THAT WAY YOU CAN GET HALF HIS THINGS."

JON: "WHOA – FOR ALL YOU KNOW I'M IN DEBT, HAVE SIX KIDS, AND A FELONY RECORD! … ON SECOND THOUGHT, COULD YOU TAKE HALF THAT CRAP OFF MY HANDS? THAT'D BE GREAT!"

Given the idea that I should strive to procure half his shit, I spent the rest of the day referring to him as my "future ex husband." I invited him to join me and my friends for some dancing at this little bar by the hotel. He declined, but I insisted. He finally agreed under the condition that he doesn't have to dance.

ME: "IF YOU MUST SIT IT OUT, I UNDERSTAND. BUT YOU'LL BE THE ONLY ONE. IT'S NOT LIKE I HAVE THE FAINTEST IDEA WHAT I'M DOING OUT THERE ANYWAY."

JON: "THAT DOESN'T MATTER, YOU'RE A GIRL."

ME: "SERIOUSLY?!? THAT'S WHAT YOU'RE GOING WITH? I HATE IT WHEN PEOPLE SIMPLIFY TO THE POINT OF PLAYING THE GIRL CARD TO GET OUT OF PROVIDING A REAL EXPLANATION."

JON: "IT'S NOT A COP OUT."

ME: "OH YEA?"

JON: "WOMEN JUST HAVE TOO MANY MOVING PARTS."

ME: "WHAT THE FUCK DOES THAT HAVE TO DO WITH ANYTHING?!? YOU HAVE JUST AS MANY MOVING PARTS WHEN YOU DANCE AS I DO."

JON: "YES, BUT YOU DON'T ASSESS THEM SINGLY. YOU KNOW HOW EVERYONE LOOKS LIKE THEY HAVE MORE RHYTHM WHEN UNDER A STROBE LIGHT?"

ME: "YEA …"

JON: "IT'S KIND OF THE SAME THING."

ME: "I DON'T FOLLOW."

JON: "WHEN MEN TRY TO WATCH A WOMAN DANCE WE GET
OVERLOADED. THERE ARE SO MANY MOVING PARTS THAT WE ONLY
LOOK AT EACH PART EVERY SO OFTEN. IT'S LIKE A BRAIN INDUCED
STROBE LIGHT. YOU ALL LOOK GOOD BECAUSE WE HAVE NO IDEA
WHAT'S ACTUALLY HAPPENING. YOU HAVE TOO MANY MOVING
PARTS!"

I simmered on this brilliant nugget for a good ten days
before deciding to put it to the ultimate test at a total dive
bar in London. I got out there, shook my groove thang, and
epileptically moved as many parts at once as possible. And
IT TOTALLY WORKED!! By the end of the night I had a
group of old timers so enamored they actually gave me a
bottle of wine to take home as a thank you for stopping by.

The whole thing was amazing, but I couldn't help but think
about how seldom such a strategy actually works in sales.
Take my favorite flub: the 'show up and throw up;' it's
arguably the demo equivalent of 'too many moving parts'
but does it net a similar result? Actually it does, but this time
it's not a good thing.

Showing a client everything, waving all parts of an
application in their face in a single meeting, will leave them
confused, not enamored. They might still think you're a
good dancer — who cleverly avoided all the tricky buttons
and ugly screens — but customers don't want to dance
around a solution. They want an actual solution.

So next time, keep it focused. Even professional jugglers will
drop a ball when they have too many in the air. When you
begin each meeting with a clear agenda you'll keep your
clients' eye on the prize, instead of the shiny object dancing
in the corner.

Product Schmoduct

Sales guys are notoriously expert deflectors. They're famous for saying things like: "I make quota because *I'm* AMAZING." But should they miss... stand back because a blame bomb is about to explode.

Having spent most of my career in the software space, I can't help but smirk when salesmen insist they've lost a deal because of — or the lack there of — a particular feature. Even moderately mature applications contain more than enough functionality to address most potential customers' needs. Success is less about what you have than it is about how you sell it.

Take my ankle for example, the story of how I broke it is lame: I leaped festively and landed fractured — that's it. Seriously. I was telling a story about how you can hear music playing from bars and shops on the street in my neighborhood, and when I jumped to demonstrate the associated joy it brings me, things went sideways.

Just a normal jump, wearing gym shoes, on dry ground, sober as a baby. But when I landed it was like my leg wasn't there. I went down. Hard. Stood back up, took a step, fell again — this time spraining it, I later came to learn — popped up again, and discovered my ankle was broken.

The only quasi interesting part is the fact that I was in London when it happened. And even that detail has left most of my audiences expecting, if not supplying, a richer series of events. Many people insist that I was drunk, others demand that I was in a fight of some sort. And the rest? The rest assume some sort of self-induced abuse.

Tired of fighting people on the facts, I decided to just start telling a version of the 'truth' more in line with their expectations. As of today I've shared this synthetic story with five lucky individuals, but two exchanges really stand out in my mind. The first, with a buzzed boy named Billy, went something like this:

BILLY ASKED (FOR THE THIRD TIME): "NO SERIOUSLY, HOW'D YOU HURT YOUR LEG?"

ME: "TO TELL YOU THE TRUTH... I DIDN'T ACTUALLY HURT IT..."

BILLY: "WHAT?"

ME: "HAVE YOU HEARD OF 'CRUTCH FOR THE CURE?'"

BILLY: "NO!"

ME: "REALLY? CRUTCH FOR THE CURE... YOU KNOW? THE 5K THEY HAVE IN MIAMI BEACH EVERY JULY..."

BILLY: "NO, I HAVEN'T, WHAT'S THAT?"

ME: "OH IT'S REALLY NEAT; IT'S TO RAISE DIABETES AWARENESS, EVERYONE DOES THE RACE ON CRUTCHES. I'VE BEEN TRAINING FOR IT FOR A COUPLE WEEKS NOW. EVEN CRUTCHED THROUGH THE AIRPORT TO BUILD UP STAMINA ... CRUTCHING THAT FAR IS REALLY HARD."

BILLY: "NO KIDDING."

ME: "REALLY MAKES YOU FEEL BAD FOR THOSE PEOPLE WHO LOSE THEIR LIMBS TO DIABETES, YA KNOW? HEY YOU SHOULD DO THE RACE WITH ME, I MEAN THERE'S STILL A FEW MONTHS TO TRAIN!"

BILLY: "YEA MY [RELATIVE] ACTUALLY LOST HIS LEG TO DIABETES ..."

(This factual curveball almost made me feel bad, but the fact he was buying this bullshit was too much fun, so I interjected before it got any deeper.)

> **ME**: "WE SHOULD GET A WHOLE TEAM TOGETHER THEN!"

> **BILLY**: "FOR SURE..."

It was at this point that he offered to recruit the student athletes at the University of Miami. Little did I know he was somehow loosely affiliated with the football team and always on the look out for charity events to involve them in. Needless to say, I had to change the subject before I dug a hole so deep I'd have to start a charity to climb out of it. To clear my conscience I suggested he google it — figuring he'd catch on — and call me the next day if he was still interested.

Hopping forward a day, I'm sitting in the doctor's office telling this story to the nurse when — about halfway through it — my doctor walks in. So I catch her up. Same deal as with Billy, I say:

> **ME**: "HAVE YOU HEARD OF 'CRUTCH FOR THE CURE'?"

Then, without even an ounce of hesitation, she goes:

> **DR. G**: "YEA."

Blew my mind, so I continued:

> **ME**: "REALLY? IT'S PRETTY OBSCURE, ARE YOU THINKING OF THE 5K EACH JULY IN SOUTH BEACH FOR DIABETES?"

> **DR. G**: "SURE, WHAT ABOUT IT?"

Hook, line, and sinker.

I'm not sure if she wasn't really paying attention, or just didn't want to seem uninformed, but since I like this lady, and my mobility is in her hands, I didn't carry on. But wow, you know?!?

Now, I'm not by any means suggesting you lie to your prospects, but the overwhelming success of this ridiculous story does raise several questions. For one, why do we assume Product Management is more gullible than our clients? The stories we tell product managers are barely plausible at times, and yet we deliver them with all the confidence and commitment as a mother sending her munchkin off to sit on the lap of a shopping mall Santa.

Yet with prospects, instead of confidently deflecting them to a more suitable solution, we shrink up. Why? There's no reason to believe prospects are less deserving of a little dazzle in the delivery than our own team is. Everyone deserves a fair shot at exiting the conversation satisfied, don't they?

So next time, go with what you got! Instead of focusing on inventing new features and functions, try inventing new uses for the features you already have. Exercising your own ingenuity leaves you sounding more genuine: the source of true conversational confidence and closed deals.

Watch Out for Swatches

Back in the day, Calvin and I would go browse furniture stores for ideas. Occasionally we get to do this alone, but more often than not we'd end up getting followed around by hopeful store employees. My favorite lost puppy salesman is named Fred and works in an eclectic little shop in Chicago.

During our last visit to the store, Calvin and I stumbled on a couch that reminded us of one we liked online. As we paused to soak it in, I asked Fred if it came in any other colors. I figured since he was hell bent on hovering and pointing out every couch in the store — without having a clue what our needs were — we might as well include him in the conversation. He said yes. But instead of following up with a question about color preference, *this* yes came with five books of fabric. Five!

Just so we're clear, I'm not the kind of girl who longs to thumb through piles of fabric swatches in search of the perfect pattern. I shudder just thinking about it. If and when Calvin and I buy a couch it'll be either black or white. Nevertheless I humored Fred and browsed his books while he lurked in the corner. But like a man who just refuses to take a hint, every time I tried to give up he'd lean in and say:

"DID YOU SEE THIS ONE?"

That's a hour of my life I will never get back.

Even though I plan to avoid ever entering that store again, I smiled and agreed to think about all of Fred's colorful ideas. Then one night I had a quasi-frightening thought! How many of our prospects sit through our demos thinking the same thing? Do our customers spend our spiel rolling their eyes, quietly hoping we shut up long enough for them to politely bow out of the conversation?

We all know you can't win them all, but I think we'd like to do whatever possible to minimize the number of wasteful demos we perform. Sure, we concede it's a good idea to ask a lot of questions during discovery, but what happens when the customer starts asking more questions than you do? Where's the line between engaged prospect and a damaging fusillade of questions? I think the answer lies in the way we field the inquisitions.

Most of us just start addressing them one-by-one. Don't feel bad, it's a reflex. When you're being bombarded with questions, especially ones you think you know the answer to, it's natural to want to answer them. At the surface this even seems like a reasonable approach. Right up until it backfires and the customer becomes worried about who, what, and how much all your answers will cost them.

What if instead you answered their questions with more questions? If Fred had just asked what color couch I wanted, I could have avoided fabric swatch overload. Simply trying to understand the source of the issue – the reason for their inquiry – enables you to deliver a specific, succinct solution to your prospect and avoid an avalanche of apprehension.

So next time, push back. When prospects attack you with questions during a demo, return the focus to their problems. By doing so, you'll maintain control of the demo, control of the sales cycle, and – ultimately – control of the close.

Genius At Work

Calvin and I were half way to a wedding before we realized we neglected to bring a gift. Luckily the reception's venue shared a parking lot with a Walmart, so I ran inside to grab a card. Going shopping at Walmart is painful enough, but to do it incredibly over dressed, in heels, in the middle of bumblefuck Wisconsin – that's a-whole-nother experience.

While in line, Calvin decided that he wanted to put fifty bucks in the card. Honestly I was surprised by the insignificance of this sum, but hey, they are his friends, not mine. Since it was bad enough the cash's value couldn't cover dinner, I figured we really shouldn't add to the tacky by taking four bills to reach it. Which is precisely how I came to try to get up-change for a fifty at Walmart.

To be clear, I had two twenties and two fives and hoped to exchange them for one crisp fifty dollar bill. Knowing that cashiers seldom have the power to open the magic money drawer on their own, I used a greeting card purchase to facilitate this high-currency access. My transaction totaled a little over four dollars and – in an attempt to lessen the complexity of the interaction – I handed the cashier $60 and instructed him to give me a fifty dollar bill as part of my change.

First he yelled at me for giving him too much money. To be fair, this wasn't entirely unexpected. I just re-explained my intentions, but he insisted that we must first complete the transaction. I conceded and took the $15 handed me, combined it with my cash in hand, and attempted to return to the man $50 in exchange for − again − a $50 bill. (Which, by the way, I could see in the drawer).

Meanwhile a woman in the next aisle caught wind of my agenda and started eyeballing me like I was trying to run some sort of a con. So now, while engaged in a stare down with a soccer mom, I continued to try to explain my consolidation quest to captain competent. This time I resorted to pointing and even suggesting with a gesticulation that I wanted the bill to go in the card I just bought. To no avail though, this guy − I kid you not − tried to hand me two *different* twenties and two *different* fives for my $50 dollars.

At this point the woman in the next aisle, who finally figured out my goal, was laughing hysterically. This drew the attention of neighboring cashiers and caused me to start laughing. By now Calvin was long gone; he doesn't find this sort of thing nearly as funny as I do. I gave it one more go.

ME: "MAY I TRADE THIS FOR A SINGLE, FIFTY DOLLAR BILL PLEASE?"

HIM: "I DON'T UNDERSTAND. YOU HAVE FIFTY DOLLARS!"

ME: "YES, BUT I'D LIKE ONE BILL, NOT FOUR."

HIM: "NO, YOU HAVE FIFTY DOLLARS. NEXT CUSTOMER."

Oh well. I walked away and glibly thanked him for trying.

The experience reminded me of the fact that the world is filled with over achievers − just like this guy. And they are often the ones in most desperate need of helpful technological solutions such as ours. Unfortunately this means we sales folk must embark on daily journeys that test the boundaries of our patience. Luckily − in theory − our product people do a great job of making the software easier and more intuitive with each new release.

But as they say, make something idiot proof and the world will make a better idiot. So how do we quickly discern between those for whom we must keep it simple, and those who will appreciate the breadth of the solutions we offer? How can we tell the men from morons?

As with most sales conundrums, this one's solution leads us back to discovery. If we know where our prospects are on our solution's adoption curve, or perhaps simply on the software definition docket, we will better understand where to start. But remember, people lie to save face, so check in often and make sure your prospects understand the value in your proposal every step of the way.

So next time, distinguish the nerds from the naive. When you tailor your pitch to the prospect at hand you will avoid unfortunate circular conversations that leave you with a fist full of potential, but still fifty bucks short of your goal.

When Teachers Attack

One hot August afternoon, I arrived in New Orleans and joined a trove of teachers for a bachelorette party. How I got invited still remains a mystery to me; these sorority sisters are a seriously girly group. But the bride finds my unfiltered commentary amusing so, in an effort to enjoy some dry wit on a humid day, she asked me to tag along. I expected a certain amount of silliness, but − having never hung out with a bunch of broads before − was wildly under-prepared for what was about to happen.

Here's the scene: it's Saturday afternoon and for some reason we're all sober. On our way back from an outdoor market, as Kim − the maid of honor − and I discussed our purchases, the other four ladies deviated course. Confused by their sudden direction shift, we hung back and watched as they B-lined it for a blonde kid who was playing in the grass. This munchkin couldn't have been more than two or three years old and appeared neither injured nor alone, so I turned to ask Kim if we knew the mother or something?

Before we could even speculate on the situation, the squealing began. This was not a new trick for these ex-sorority girls, and I still honestly don't get it; life is exciting − sure − but there's no reason to yell about it ALL THE TIME. Seriously, I'm *right* here, I can hear you!

But, I digress ...

Kim and I just stood there, jaws dropped as they squeal-giggle-clap-danced all the way up to the child. A stranger's child. Totally bypassed the mom. Didn't even do the parent the courtesy of a hello. Nope, these four grown women surrounded this little girl, for no other reason than: "she's so cute," and they "wanted to play" with her. When one of them pulled out a camera as the others posed for a picture, my confusion turned to disgust.

"IT'S NOT A PUPPY!" I SAID AS I TURNED MY ATTENTION TO KIM, "WHY ARE THEY PETTING IT?!? YOU KNOW, IT'S SHIT LIKE THIS THAT GETS YOU ARRESTED."

The mother was as confused as I; when she tried to break up this ban of smiley child predators, one of the girls turned to her, laughed, and said:

"OH, HI! IT'S OKAY, WE'RE TEACHERS..."

She then promptly resumed the fawning. The mom − in shock − just stood there. While Kim walked over to move the madness along, I couldn't help but wonder if we apply similarly lame excuses as we sell.

Are we so used to our product's problems that we dismiss legitimate client concerns? Or worse, do we focus so tightly on a single solution that we fail to notice how silly it seems? When trying to find balance between processing the pipeline and nursing new interest, how do we ensure the right approach?

Simple. Don't cut the line.

Just like in grammar school where 'cutters' were shunned by the class, salesmen who bypass the chain of access will face frustration in the future. Sure, you might want to go directly to the decision maker − or run right up to a random toddler − and it might even work out in the short term. But eventually, eventually you'll find yourself face to face with their assistant (or mom) begging for continued access, and I doubt they'll care to consent.

So next time, acknowledge all parties early in the process. When you have sufficient confidence in your agenda to share it − even with those who might not approve − you'll find your future efforts far less futile.

Free By Definition

While browsing my newsletters email account the other day I found a story about people who pursued a free tattoo[5]. The article revealed that 68% of participants would not have gone through with it if they had been asked to pay. That's right — two thirds of people are willing to permanently augment their body if, and only if, it doesn't cost them anything.

The way I see it, 'free' falls into three categories: stuff I would do, but I'm too cheap to pay for — free from financial obligation; stuff I'd never think or want to do, but will because it's available and doesn't cost me anything — opportunistic freedom; and stuff that only appears to be free — you know, scams.

This reminded me of a well known and accepted marketing tactic: 'free gift with purchase.' Clearly this falls into the middle division, but what category do software promotions call home? We often offer 'this for the price of that' or 'an add-on for no extra charge,' but are we assuming these items are otherwise coveted and their acquirers simply lack budget? Or are we inadvertently wooing those who will try anything once, as long as it's free?

In my experience, the warning signs appear early on. For example, I once attended a jewelry party[6] where you could get an additional ten bucks off if you "schedule a time to hear about the opportunity to own your own business."

---

[5] HTTP://DANARIELY.COM/2010/11/10/THE-POWER-OF-FREE-TATTOOS/
[6] THINK TUPPERWARE, BUT JEWERLY.

I like the idea of a free $10, so I walked up to the lady and said:

"I SHOULD TELL YOU, I'M REALLY GOOD AT SAYING NO, BUT I'D LIKE TO HEAR WHAT I NEED TO DO TO QUALIFY FOR THE COUPON."

Now, I don't know about you, but if someone said that to me, I'd ask one, maybe two, questions to qualify their (dis)interest and probably just concede the discount for the sake of time. But OH NO! This woman went into full-on hard sell mode.

After twenty minutes of me systematically tearing down her arguments with remarks like:

"YOU'RE LOOKING AT BOTH MY FEMALE FRIENDS."

"NO, I DON'T HAVE AN OFFICE OF WOMEN TO SELL TO."

"NOR DO MY MALES FRIENDS HAVE WIVES WHO NEED CHEAP JEWELRY."

"AND I DON'T ATTEND A SEWING CIRCLE, BRIDGE GROUP, OR AEROBICS CLASS EITHER THANK-YOU-VERY-MUCH!"

… the fun had faded and I bowed out of the conversation by reminding her that I wasn't really interested before ultimately thanking her for the coupon.

So next time, don't mistake interest in a promotion as genuine interest in a solution. Tattoos might last forever, but remember: most products are replaceable and the return on investment for both vendor and customer continues long after the promotion concludes; it's your responsibility to sell the solution's value, not hawk an alleged bargain.

Stereotypically Sad Service

Over the course of a single month I had five separate conversations with my lovely cable provider. It all started when I called to cancel the TV portion of my cable service; I explained that the conclusion of football season coupled with the prevalence of television streaming services had obsoleted that portion of their service offering and I would only be needing Internet going forward. They requested I return the cable box to a facility and I agreed to comply just as soon as I returned from a business trip to Asia. What followed was a series of phone calls, each furthering my rage and hatred of Comcast.

Just days later, while I was in Singapore, they rang to ask when I planned to return the equipment − I advised them to review their notes and, again, to expect it on the 5th. On the 13th their sales department called to ask if I'd like to bundle my stand alone Internet services with television to save $15. I asked what about me cancelling TV service not three weeks prior suggested I'd be interested in buying it back today? The rep failed to appreciate the sardonicism and quickly disengaged.

Then, Friday the 20th, while waiting for a taxi at O'Hare[7], I get a call. This time it's someone from billing calling to inform me that they suspended my service earlier in the week and wondered why I hadn't called in to inquire about it. I applauded their passive aggression before asking how my account could possibly be past due when they've sent me emails each and every month indicating my enrollment

---

[7] O'HARE INTERNATIONAL AIRPORT (ORD), CHICAGO, IL

in automatic billing. She had no answer, so I pressed on:

> "HOW IS IT THAT I'VE SPOKEN TO [COMCAST] THREE SEPARATE TIMES THIS MONTH AND NOBODY MENTIONED MY ACCOUNT WAS IN ARREARS? YOU KNOW THEY MAKE SOFTWARE FOR THIS RIGHT?!?"

I went on to explain that if she thought I was going to pay a $7.28 fee for *their* incompetence that she was out of her bloody mind. I sooner cancel the rest of my service right then and there. The conversation almost ended with me settling my tab and moving to AT&T, but the agent advised me to call back on May 7th to confirm auto payment receipt.

> "YOU'RE JOKING RIGHT!?!"

She wasn't.

> "LOOK THAT'S NOT GOING TO HAPPEN. WHY DON'T YOU PUT A NOTE IN MY ACCOUNT, OR SET A TICKLER FOR SOMEONE TO CALL ME THAT DAY OR SOMETHING. THEN WISH ME WELL AND CELEBRATE THE ONSET OF YOUR COMPETENCE. SERIOUSLY IF Y'ALL CAN'T HANDLE AUTO-BILLING THEN YOU SHOULDN'T OFFER IT AS A SERVICE. EITHER WAY MY INVOLVEMENT IN THIS MATTER REALLY MUST DRAW TO A CLOSE."

The agent used the final few moments of the conversation to defend her point and I hung up laughing after she broke into the standard 'thank you for calling Comcast' close. Apparently after all that she forgot who called whom.

Which brings us to the end of the month.

I again get a call, this time from sales — who apparently felt that the only way to respond to a near firing is with an upsell. Luckily this agent properly registered the tone in my "Comcast! What can I do for you today?!?" greeting and didn't try too hard to convince me that doubling my internet speed was invaluable. She actually skipped the thank you and just fired off a "1-800-COMCAST" and hung up.

Clearly they lack — and desperately need — CRM. Companies with proper visibility into the interactions they have with their customers would never behave so schizophrenically. This kind of siloed behavior only leads to resentful customers who complain publicly. Luckily — or unluckily perhaps — they'll never appreciate the risk they introduce because I'm sure they also neglect to monitor and map complaints on social channels back to their customer rolodex.

So next time, consider this a cautionary tale. When you don't update your CRM you're only hurting yourself. Those five seconds you think you're saving by neglecting to add notes from your call — as we can see — quickly turns into hours of wasted time and effort. Focus your funnel; document your efforts; hit your number.

Living the Infomercial

My friend Tammy was in town the other day for a work conference and agreed to meet Josh, Shale, and me at a bar called Finnegan's Road. Josh — a random dude I met while drinking at one of my go-to spots in South Beach — hit happy hour pretty hard and desperately needed to eat something. So when Shale picked the two of us up, I instructed him to stop for some 'Crack Chicken' on the way. This delicious detour, however, made us late.

We arrived at Finnegan's about twenty minutes later than planned. So after a half a cocktail and still no sign of Tammy, I began to wonder — okay, worry — about where the hell she was. Last I heard she was wrapping up dinner not two blocks up the road; there was no way I should have to beat her to the bar, let alone have the time to consume a cocktail before she arrived. My concern must have made her ears ring because not a moment later I received a text:

"COME RESCUE ME'. I'M AT ADORE."

The problem was I had no idea where this 'Adore' bar was. So, rather than wait for a return text, I called her. Turns out it wasn't a bar at all; Adore sold makeup and is conveniently located two storefronts away from Finnegan's. Figuring this shouldn't take too long, I left my cocktail on the counter and told the boys I'd be right back.

I walk into this brightly lit, barely occupied store and find Tammy sitting in a beauty chair, flanked by two sales reps, clutching a half empty bottle of Jack Daniels loosely wrapped in a Mango's Tropical Cafe to-go bag — apparently her business trip proved more festive than expected. The scene was something out of an intervention reality show. If you didn't know better you would have thought these two clerks were trying to convince poor Tammy that lotion is a suitable substitute for alcohol.

As I approached Tammy the sales chick greeted me with:

"AH, YES. WELCOME. NOW TELL ME, WHICH SIDE OF HER FACE LOOKS MORE REFINED?"

Having no idea what constitutes "more refined," I hesitated. The guy glanced my way and tried to signal to his partner that I was drunk. Trust me, I wasn't *that* drunk. Plus, even if I had been, between the chicken and the challenge I was immediately, officially 1-2-3 sober.

Still not sure whether Tammy wanted me there to negotiate a discount or stage a prison break, I started slowly. In an overly pedantic fashion, I explained how any discrepancies between the sides of her face were simply too subtle for my astigmatism laden eyes to detect. I then explained how the ice was melting in my cocktail next door and that Tammy and I should really get back to the boys.

This remark relocated us to the counter; they mistakenly believed Tammy — who'd been giggling the whole time — was ready to check out. We would have been out of there sooner, but Tammy kept undermining my retorts with facts. Little miss "oh no Erin, my flight is in the morning, we don't have time to come back tomorrow" and "no, the bottles appear to be under three ounces" forced me to get creative.

I split up the team by suggesting the gentlemen join me for shot of the Jack, hoping Tammy could handle the lady one on one. When we returned the guy knew it was now or never. He cleared all but one of the items away, and said:

"[GIRL'S NAME] JUST GIVE IT TO HER 50% OFF."

What happened next was amazing. The chick did the 'fifty?!? You're gonna give her *50% off??*' fake-shock dance – complete with quick head turns between him, us, and the register. Then he's like:

"YUP FIFTY. THEY ARE SERIOUS AND I DON'T WANT TAMMY TO MISS THIS OPPORTUNITY."

I could barely keep from laughing; this was like something off bad late-night television. I was ready to go and they knew it, but before I could even turn to leave Tammy decided to go and ask:

"WHAT ALL CAN I GET FOR THE 50% OFF."

"LOOK!" I EXCLAIMED BEFORE ANYONE COULD RESPOND TO HER INQUIRY, "50% OFF ISN'T A PRICE, IT'S A DISCOUNT AND I BET NO ONE HAS TOLD YOU HOW MUCH THE RETAIL PRICE ON ANY OF THIS SHIT IS YET, RIGHT?"

They hadn't. I demanded the price. MSRP: two THOUSAND dollars! I looked at Tammy, told her there was no way I was letting her spend a thousand dollars on lotion, grabbed her with one hand, the bottle of Jack with the other, and left.

So next time, remember this *is* real life. I understand pricing isn't as transparent as it used to be; customers expect discounts, but there's a fine line between meeting budgetary requirements and cheesy infomercial guy. When you can establish real value you'll never have to holler "but wait there's more" as they walk out the door.

Neighborhood Bars

Despite the constant heckling of my colleagues, I make no secret of — nor apologies for — the fact that my preferred hotel group is Intercontinental. Proudly, I generally opt for Holiday Inn brands. So it should come as no surprise that my favorite place to stay near Cupertino, California is the Holiday Inn Express on El Camino Real. It satisfies my hotel selection criteria: free food, free internet, firm pillows, a shower head that doesn't attack your nipples, and a housekeeping staff that's okay with *not* cleaning my room for a week at a time.

While I'm pretty easy to please, such simple luxuries don't come without their share of compromise; this hotel in particular is located in a neighborhood that belongs in a Chris Rock bit. The drive up goes something like: liquor store, porn store, liquor store, shady motel, porn store, strip club, Holiday Inn Express. So you can imagine the adventure when Chip and Ned — a couple colleagues also in from out of town — decided to join me on a hunt for a bar we could walk to.

After some discussion we decided to try the place right across the street. Upon entering we quickly suspected they sold more than alcohol at this bar. We perched ourselves on stools and watched the parade of scantily clad lady-types promenade in and out of the back room at interestingly timed intervals. With such a variety of women available, the beer selection was surprisingly horrible; so I ordered my standard:

ME: "RUM AND SODA PLEASE."

BARTENDER: "WE DON'T HAVE THAT."

I assumed she meant club soda.

ME: "NO BIG DEAL, I'LL JUST TAKE RUM AND WATER. WITH A LIME, IF YOU HAVE IT."

BARTENDER SMILING WIDELY: "WE DON'T HAVE THAT EITHER."

ME: "WHAT? YOU DON'T HAVE RUM? OF ANY VARIETY?"

BARTENDER: "NO."

I could see bottles of Ketel One so I changed my order to a Kettle One and water. The boys went on to order a couple Jack and cokes. Simple enough.

The bartender left, presumably to assemble our cocktails. When she returned she served me a simple shot glass of – what I assumed was – vodka. Chip jumped in, attempting to correct the situation. With a series of very targeted gesticulations he attempted to communicate the booze in a glass of ice concept.

The bartender nodded and went back, but again returned with shots in dark colored shot glasses. We couldn't assess the translucency of our cocktails, so we had to trust her assigned placement. We thanked her and – again – tried to communicate our desire for ice and mixer.

In a third round trip to the far side of the bar, she managed to bring us three chasers. We rolled with it. After knocking back our shots, however, we became confused; none of the booze tasted as expected. The bill revealed the boys were drinking Patron and I got some kind of Chivas.

Apparently tequila is to whisky what vodka is to scotch.

The whole experience was such a hot mess that all we could do was laugh. But it made me wonder how often sales people simply disregard the customer's request in favor of pushing their own agenda?

Since we didn't identify the issue until after we consumed the product — in a single swallow, per her clever design — we missed the opportunity to request a corrective action. Did that somehow make her mistake okay? Does the 'better to seek forgiveness than ask permission' mantra apply to sales situations like these as her behavior would have us believe? Can a 'bait and switch' ever be acceptable?

Ten years ago, maybe. I mean sure, her alcohol choices — in theory — delivered the same result: we got one ounce closer to a proper buzz. But today social media gives every loudmouth with a keyboard a voice. Suddenly you're not just asking a single wronged customer for forgiveness. Now everyone knows of your troubles and demands your apology. These days failures are touted so much louder than successes that a company's customer service defines their brand as much, if not more, than their product.

So next time, sell honorably. The expectations you set are as important as the product you deliver; set your implementation team up for success by being true to your customer and true to your product. By doing so, you'll ensure everyone gets exactly what they ordered.

Right on Target

Ralph took me shooting one weekend. We'd been planning to go for a while now; during which time, I've been teasing him about how it's about time my favorite fed teaches me to shoot properly. By the afternoon of the event, however, I realized that I might not live up to his expectations...

People often mistaken my *mental_floss*[8] driven knowledge of subjects for genuine expertise. It happens with motorcycles all the time. Sure, I've had my motorcycle license for seven some-odd years, and yes, I've only ever owned sports bikes. But that doesn't mean I'm a riding zealot. Far from it. I've got chicken strips for days, wouldn't dare lane split, always wear a helmet, and — while I may have once tried to clean my carbs with my buddy Ajax — couldn't tell a spark plug from a piston if you paid me.

It's kind of the same thing with guns. Including this weekend, I've been shooting all of four times. Inexperience aside, I managed to enjoy a conversation about firearms with a half a dozen retired federal agents the Saturday before. I didn't lie or misrepresent my specific knowledge at all, but — as it is with the bikes — boys hear I have a concealed carry permit, brand me as "one cool chick," and start talking to me like I'm a resident expert.

I don't mind. It's nice to be included. But it did get me thinking, do we jump to similar conclusions with our clients? Today's customer is more educated than ever. Many know more about our product before they call than we did three

[8] MENTAL_FLOSS IS A MAGAZINE, CHECK IT OUT AT MENTALFLOSS.COM

months into selling it. So how can we tell who's really got the topic in their sights and who just has the lingo down?

Take a lesson from Ralph – ask. When we got to the range, before diving into a lesson or assuming I knew it all, he looked at me and said:

"OKAY, SO SHOW ME HOW YOU SHOOT."

He gave me the opportunity to demonstrate my current level of knowledge, and himself the opportunity to learn about me before he pitched me with a lesson. Now, I don't know if it was nerves, naivete, or simply beginners luck, but my first shot was SPOT ON!

This probably happens in sales cycles too. You ask the customer an opening question, and it just so happens to be the only question they know the answer to. Don't mistake this single successful shot as straight up expertise. Ralph sure didn't. And even if he had, my next six rounds revealed my true colors.

By allowing me to establish the depth of our impending conversation, Ralph was able to impart onto me some very actionable advice. Now, I'm not suggesting you demand your customers debrief you on your own product to gauge their competence. Instead use your savvy sales skills to engage them in a conversation that subtly reveals at which expertise level they reside.

So next time, go beyond the sound bite. A comment or two doesn't mean they know as much as you. (If they did, they'd have bought already.) Take your time; let them guide you. When you do, you'll gain the perspective necessary to keep your team on target and the close in your sights.

Teamwork

They say humans are social creatures that evolved both to and because of their cooperation abilities. Yet historically our most famous leaders are reknown for their subpar interpersonal skills. Evidence suggests that Beethoven was bi-polar, Darwin was delirious, and Dickens was depressed. True to geek stereotype I, too, tend toward the inept side of the social spectrum from time to time. While I've adapted to my shortcomings by developing an overgrown sense of self-reliance, I – like even the most independant specimens of our species – occasionally need a little backup.

As was the case the other night when I decided my time as a motor-free Miamian should come to an end. So I scoured Craigslist and found a suitable motorcycle specimen. In my experience, however, boys don't enjoy it when girls haggle for their toys. So I solicited the negotiation – and bodyguard – services of my friend Tuck, a local business owner. He's tall, strong, and presumably business savvy – but more importantly, Tuck's an alpha.

As one myself, I often struggle to stick to my part in a good-cop, bad-cop negotiation. In an effort to stay on script, we decided ahead of time what I would pay and outlined a strategy based on details available from the ad, Internet, and photos. After a quick inspection of the bike and lap about the library lot, the negotiations officially commenced.

I opened with my final price, mostly because I get some sick joy from getting other people to move without moving myself, but also because I see wavering as the weakness that woos overpayment. As the haggling progressed I became so amused by how well we were doing unrehearsed that it became difficult to maintain my poker face. At one point Tuck got a call. As he looked at his phone to check the caller ID, I go:

"IS THAT THE OTHER GUY CALLING YOU BACK?"

TUCK: "YEA, IT'S THE GUY FROM OUT IN PINES."

Our 'decision' to send that call to voicemail until we saw how this deal shook out became the closing force we needed; despite still being $500 apart, the seller succumbed to our will.

As I talked the bloke through the paperwork — basking in the glow of my victory — I couldn't help but wonder why sales guys have such a hard time sticking to sticker price. I mean why, oh why, do we apologize for the cost of our product? Do we simply not understand its true value? Are discount friendly compensation plans to blame? Or do we just suck at delivering a compelling argument?

Whatever the reason, since software has gone the way of automobiles in its pricing, we've gone soft! It's sad really. Just as your customer knows what they are willing to pay, you should know what you're willing to charge. When we volunteer facts about pricing flexibility, all the buyer hears is: "whatever the quote says... it's too much!"

So next time, stand your ground. Applying this tactic netted me a new bike and even scored me a loaner license plate. When you deliver your contract with enough conviction to compel prospects to your position, you'll save your quids for bigger gives, and ultimately close like a pro.

A Carolina Rabbit Hole

You know those restaurants that try to upsell you on water? Places where the waiter begins by asking you whether you prefer your water sparkling, still, or perhaps glacial? Evidently it's the polar bears' tears what give it the great flavor! Well, I had the pleasure of dining at a couple such fine establishments recently, and the assortment of specials at one in particular inspired me.

Our server delighted us with very detailed stories of trout served upon a bed of potatoes with a reduction of some sort, and rabbit farm raised in Carolina roasted alongside seasonal squashes. I hadn't previously realized that rabbit, like fruit, varied by region and upbringing. So since location clearly mattered, when the waiter paused for questions, I casually asked:

"WHICH CAROLINA?"

The table — half filled with folks accustomed to my sense of humor and the other half a collection of fresh-on-the-team business partners — met the remark with a chuckle, while the waiter took pause. I stumped him. He was clearly uncomfortable when he asked to be excused to research my question. A few minutes later he returned to inform us these rabbits originated in South Carolina. I looked around the table and asked:

"IS THAT THE BETTER CAROLINA FOR RABBITS?"

As I let the waiter off the hook and placed my order, I couldn't help but be reminded of a demo folly many sales folks frequent: the feature spill.

I understand that in the world of food, adjectives play a key role, but certain details invite more questions than they add value. So, as salesmen, how do we stick to the details that aid in the decision making process and dismiss the rest? Is it possible to make your product sound fancy without filling your story with fluff?

Sure it is! In the case of a demo, simply remember — your prospects can read. As you gesticulate towards the features of your solution speak only to their merits not their physical attributes. Despite what you might think, a feature's feature is not a benefit.

So next time, before you wax poetic, ask yourself: "so what?" Sure, you might get away with inviting irrelevant questions, but every so often you'll find yourself face to face with a prospect like me — someone who will ask, just to ask. Just remember, when there is no right or wrong answer, it's easy to get caught in a Carolina rabbit hole.

The Perspicacious Paratrooper

Shale called one afternoon to accuse me of exiting his truck "like a paratrooper" the night before. I have no specific memory of that precise moment as it was four o'clock in the morning and I was appropriately inebriated at the time. But given the height of the vehicle, the whimsical architecture of my skirt and − well − myself, I could see that.

You see, I've spent the last decade honing my autopilot procedures. Sure, I used to occasionally get on the Edens expressway in Chicago heading the wrong way − a vestigial side effect of my orthodontist's office's location − but I've since compensated. Today my autopilot protocols protect me from the two major forms of social solecism: impending memory malfunction and scatterbrained salesman. In defence of either episode, however, the goal remains the same: buy time while I dial back in to reality.

Take one of my discovery calls from earlier in the week, while wrapped up in a sidebar instant message conversation, I suddenly realized it was my turn to talk. The problem was, however, I totally failed to assimilate the customer's last few lines. Rather than risk false commentary, I deployed Captain Nonsequitur. His rescues come in the form of universal truths and tangential inquiries.

In business, despite what people want to believe, everyone is *not* a unique little snowflake − they just wish they were. To capitalize on this fact I simply replace assertions with questions − and create an opportunity to play off the scatterbrained moment.

I asked a couple of questions regarding the future big picture direction of the company, examined whether the gentleman had considered a few boilerplate customer relationship strategies, and touted us as the vendor who's interested in the long game just as much as the success of the block-and-tackle brigade. I concluded with some apothegm about customer engagement and handed the reigns back to the salesman. I got what I needed; I understood the prospect's vision, planted seeds to position myself as a thought leader, and − most importantly − got him to confirm these facts, unsolicitedly.

What happened next however, really made me wonder why others struggle to capitalize on conversational serendipity. The customer was so impressed with our grasp of the big picture, he asked me how to proceed. As an SE, however, I didn't want to exclude the salesman so I called on him for collaboration. Unfortunately, he froze!

As I took back the reigns, I got to thinking about sales scripts. What makes us momentarily forget all our dance moves when the cadence of a conversation makes a sudden adjustment? Why don't we have a sales autopilot protocol in place to handle easy deals? When a customer gets excited, so excited that he concedes control of his buying cycle to you with a bow and a bow − accept the bloody present!

So next time, work on your uncertainty protocols. With your sights set on the endgame − whether it be home safe before a blackout, inspiring people to think beyond software features, or simply an expeditious sales cycle − you'll stop getting in your own way, and will find much more success in the future.

Throw a Dart

Alaska Airlines sent me an email to remind me of an outstanding voucher I had with them that was about to expire. The problem is, Alaska doesn't fly to MIA. In fact, the only service they provide to South Florida is a single flight between Fort Lauderdale and Seattle. Lauderdale is far, and I've already been to Seattle a few times, so I decided to explore alternatives.

My first thought was to catch an Alaska flight from somewhere else in the country, since that's how I came to have this voucher in the first place. This proved more expensive than it was worth, literally. Then I decided to look into codeshares[9]. Turns out Alaska codeshares with American Airlines between MIA and both SFO[10] and LAX. Again I've been to both those places a bunch of times, but in the spirit of spending free money, they'd have to do.

Rather than fly across the country for a weekend to revisit a city solo, I decided I'd try to convince my best friend to join me. Ralph has never been to LA, so he thought this was a delightful idea. We picked a weekend and − first thing the next afternoon − I headed to AlaskaAir.com to book.

But Alaska vetoed my plan. Apparently you can only use vouchers on Alaska metal; codeshares don't count. Frustrated, I returned my attention to Google Flights hoping that perhaps I overlooked a plausible place to go. I hadn't.

---

[9] A CODESHARE IS WHEN YOU FLY ON ONE AIRLINE'S FLIGHT NUMBER ON ANOTHER AIRLINE'S PLANE.
[10] SAN FRANCISCO INTERNATIONAL AIRPORT

But I had overlooked an incredible deal — round trip tickets between Miami and Bogota, Colombia for $202 per person! So I texted Ralph:

"HEY, IF WE SIGN YOU UP FOR A PASSPORT, HOW WOULD YOU FEEL ABOUT GOING TO BOGOTA INSTEAD OF LA?"

Much to my delight he responded:

"I'M IN!"

Anyone who responds to a request to spend fortynine hours in Colombia without question, hesitation, or concern is absolutely my kind of person. But as I booked our trip I couldn't help but think about how rare such free spirited and flexible friends are. Prospects — or anyone who doesn't know you personally, for that matter — are probably even less likely to go along with any hyper-revised plan you present.

What happens when you're about to close and something unexpected comes up? It happens more often than we probably care to admit; something changed in support, your sales engineer accidentally overlooked a deal breaker, or legal no longer likes the terms — none are your fault directly, but all are your problem to solve. How do you break the news without breaking your momentum?

Simple. Always deliver solutions, never problems.

So next time, be ready with plan B. On the surface it may seem like I threw a dart and flew us to Bogota, but in reality, I preserved a mini-break and saved us $300 in a single detour. Perhaps with the right spin, your customer will discover similar savings in the alternate agenda. Either way, only when you have a plan, do you have a chance to close.

Boxanne on a Budget

I decided I wanted a car the other day. Not need, I still don't *need* one (for the record), but if I was going to get back into bowling, I really should procure weather-friendly wheels. Don't get me wrong, I considered bungying bowling balls to the back of my motorcycle, but ultimately decided a proper car was a better plan. So off I went, in a rental, up to Hollywood Toyota where I proudly marched into the showroom to look at some cars.

On the way in I passed upwards of a dozen dudes, presumingly sales reps — but you wouldn't know it from their reaction. I scurried on past without so much as peep from the peanut gallery. Inside, much to my dismay, there were no cars... many more men, but no motor vehicles. Not a single one. And I looked!

I spent the next eight minutes casing the joint; I went up and down every corridor, scoped out service, tested the toilets, even passed by parts — nothing, no cars, no howdy, no pitch, no assistance.

I guess when there's no weather, you don't keep cars in the 'showroom.' I returned to the lot, but even though confidence was high that these were the cars I was supposed to browse, something still felt wrong about poking about a parking lot all alone. Plus I already knew — within two — what kind I was going to get. So I sauntered over to scout the Scion building. Perhaps there I'd find a proper salesman.

Again I enter the room with conviction, this time to find six 'sales' guys and one car. A marginal improvement over the Toyota tent. After walking the entire building, just as I was about to give up, right as I began to actually laugh out loud, a head popped out from behind a computer monitor.

BOB E. BOBÉ: "HI."

ME: "DUDE! YOU GUYS GOT TO BE THE LEAST AGGRESSIVE SALES TEAM ON THE PLANET!"

After filling the man − with the best name EVER − in on the details of the last fifteen minutes he said:

"OKAY, LET'S GO LOOK AT SOME CARS THEN."

Like I said I had already narrowed my selection to two; I wanted either a Scion xD or a Rav4. After realizing the xD was a toy car, the Rav switched trunk styles in 2013, and the xB had the same engine as the Rav, I updated my scouting plan to include the xB. Much to my delight, they had a suitable specimen in stock. With my deal-breaker criteria met − Toyota, brand new, and blue − I was ready to move onto the fun portion of the evening: negotiation.

During the tour Bobby kept showcasing 'clever' features − like keyless start and USB ports − that I neither needed or wanted. Frankly I'd forgo anti-lock brakes if I could; I'm a proper Chicagoan − I know how to brake. But just because it is hard to find cars without such 'amenities' these days, that doesn't mean I feel the need to pay for them.

This was the basis of my position.

I had them run my credit and what not − just to keep things interesting − before ultimately offering them something in the neighborhood of $900 under dealer invoice. (Sticker price never even entered the conversation.) While management considered my offer, Bob and I went to pick up Boxanne — as my girl would later come to be called — so she and I could get further acquainted.

I returned, not surprisingly, to a counter offer — which I, of course, countered. By now my mother had come back with invoice prices for the same model at other south Florida dealerships that were lower than Boxanne's. I presented the evidence to Mr. Bobé who explained that Boxanne came with many additional features: Toyotacare, tinted windows, etc. But remember, I don't feel the need to pay for things I didn't ask for — even if they are useful. So I stuck to my position: I'd pay the invoice price of the car with only the features I wanted.

Rather than letting me patronize the competition, they accepted my logic. We landed at ~$400 under invoice.

As Bob, the friend of his we suckered into schlepping the rental, and I caravanned our way back to the beach, I couldn't help but think about all the boys who passed up on the opportunity. I mean sure, I wasn't dressed swankily and there was no evidence of a flash roll in my pocket, but you have to wonder how often someone saunters into a dealership, solo on a Sunday night, without buying intentions?

So next time, say hi! Part of maintaining a healthy pipeline is continually topping it up. If you can forgive the cliché, remember — while not every hello will hand you a close, surely none of the ones you pass on can. Hedge your bets salesman, the bluebirds are calling.

Assumptions

New car notwithstanding, I decided that if I was truly serious about getting back into bowling, I should get a new bowling ball. So I ordered myself a new ball online and took it down to the proshop at Bird Bowl in Miami to get it drilled. I headed to the bowling alley with two agendas: bowl (duh) and scout a league for next season, a.k.a. make friends.

While schmoozing in the pro-shop I handed out two business cards: one to an older bloke who said his team may need a bowler next year, and the other to the good-looking twenty-something proshop dude who said something about meeting to practice on Sundays.

The next morning I awaken to a text:

"HEY IT'S EDDIE FROM THE PROSHOP."

I − remembering the pro-shop guy's name started with an E, or at least an E-sound − immediately jumped to the conclusion that I was chatting with him. (I later learned his name is actually Esteban, not Eddie.) What's worse, I landed so firmly on the assumptive fact that I missed clues to the contrary along the way. I didn't notice when he referred to proshop employees in the third person and I managed to ignore the peculiarity of his suggestion to hang out "right after work," as if the traffic between us at that time of day isn't horrendous.

So, oblivious to the flags, I made plans to meet up.

Thankfully I'm a cyber stalker, and I always demand − and

subsequently validate — last names. Turns out Eddie is a bit of a net junky himself; he made it easy for me by showing up as a 'recently viewed [my] profile' person on LinkedIn. The thing is, Eddie graduated college in 1986. Now, while I SUCK at gauging people's ages, I pegged Mr. Proshop as being younger than me, which means at worst he's forty.

Now there I was, committed to an accidental date with a dude whose intentions I had not yet taken the time to preemptively align with my own. I ultimately got out of it with some lame, albeit sort of truthful, excuse about having other plans. But not without wondering how I let my guard fall so far that I failed to properly qualify the proposal.

As sales engineers, how often do we get so excited about a client asking a question that we fire off elaborate answers before they even reach the punctuation? As salesman, aren't we sometimes a little too eager to share how another client overcame the 'same challenge' before first confirming this lead indeed struggles similarly?

Sometimes we just need to know when to slow down. When handing out business cards, go-go-go. But when qualifying client calls, pause and know when to say no. Networking and prospecting may go hand in hand, but they aren't twins. Different skills are required for each.

So next time, read what's on the lines, not just between them. Let your prospect finish sharing their problems before you try to solve them. When you ask the confirming question, even though you *totally* know the answer, you might just avoid committing to a concession — or a coffee — you never needed to give.

Cash Only

The general consensus among my European colleagues seems to be that Americans arrogantly blow through the region with an unfounded confidence expecting everyone else to adapt. On my flight over the assessment baffled me, but after four days in Munich, I think I get it. You see − in addition to being big fans of rules − the German culture is very anti credit card. When you pair this fact with the weakness of the Dollar against the Euro, you get an influx of seasoned American business travelers arriving with their pockets packed with cash.

Seriously. Normally, most Americans barely carry enough cash to pay their friends back for lunch. So to deplane with funds sufficient to sustain a week abroad − especially after being killed by the exchange rate − leaves even the most humble traveler debarking with the swagger of a rapper on their way to the strip club. *We's be ballin' baby!*

I don't bring this up to encourage the targeting of tourists for theft or anything, but rather because − as my pockets deplete and the swagger subsides − I began to wonder how these cultural differences impact negotiations. Do countries that subsist on credit expect different payment terms than those who usually pay up-front? Or more importantly does the immediate bottom line impact that comes from a real-time withdrawal impact the amount of 'proof' you require prior to purchase?

Americans, conditioned by infomercials, expect things like a 'thirty-day, money-back guarantee' and, as such, are generally willing to take a leap of faith with a product. Yet as my German colleague Piers tells me:

"THAT KIND OF FOO FOO SELLING WOULD NEVER WORK WITH THE PRAGMATIC GERMAN SHOPPER."

My contacts in Asia tell a similar story, citing proven features as the primary selection criteria.

I'd like to believe we can overcome this skepticism by seeking a balance between vision and reality. The perceived polarization between theory and practice — between features and returns, between proof and faith — has long been a challenging void to span. Yet I believe — quite strongly — that without a goal, the journey lacks purpose and the features don't matter.

Especially in the software subscription space, starting with the end — the goal — in mind creates differentiation and gives you the head start necessary to overtake the competition. Just like the returns you promise your clients, this early lead can be maintained year over year. But all that time you spend proving you're not selling, and you're letting the competition catch up. If your client expects you to have faith that they will buy after you build something, they should be willing to have some faith too — with or without sixty-day payment terms.

So next time, respect cultural subtleties but don't allow diversity to derail best practices. Strive to strike a balance between what you know to be best and what the customer demands. When you do, you'll effectively guide the buyer, cash in hand, across the finish line.

Enhanced Femininity

The more I am asked to write stories that tie the features of an application back to the impact positive change will have on a customer's business, the more I spot 'epics[11]' in everyday life. In fact I find using personal motivations, and the methods by which we identify them, to be a very effective sales training tool. Seeking to pinpoint one's personal buying motivation — their epic, their goal — carries with it a clear value in countless commercial applications.

Why, just the other day I found myself faced with an epic inquiry at — believe it or not — the doctor's office! Immediately beneath the standard, feature-centric question: "What is the purpose of your visit?" laid discovery gold! This doctor made a point of asking: "What is your goal with [this] surgery?"

I became so instantly excited by the questionnaire, I actually had a hard time answering the question. I thought:

"WHAT'S MY GOAL? WELL, BIGGER …"

"NO, THAT'S NOT AN EPIC."

"RENDER THE YOUNG MEN'S DRESS SHIRT DEPARTMENT OBSOLETE …"

"URG! THAT'S NOT AN EPIC EITHER."

---

[11] 'EPIC' IS A TERM COMMONLY USED IN CERTAIN COMPUTER PROGRAMMING METHODOLOGIES. FALLING UNDER THE UMBRELLA OF A 'THEME,' IT'S MEANT TO DEFINE A BUSINESS DRIVER BEHIND A SOFTWARE'S APPLICATION.
IN THE CONTEXT OF SALES, I USE 'EPIC' TO MEAN THE SPECIFIC KEY PERFORMANCE INDICATOR (KPI) A CUSTOMER HOPES TO IMPROVE WITH THE PURCHASE.

This frustration snowball precipitated an avalanche of paranoia. What if my answer is lame, or bad, or wrong?!? Will the doctor elect to cancel my consultation? I realize no motivation, personal or professional, can truly be wrong per say; just like with any quality discovery question, it simply sought to exclude customers whose goals require a toolkit outside the doctor's skill set.

To quiet my mind, I jokingly asked Ajax — who I brought along to provide his famously objective opinion — what he thought. While we tossed around some comically bad ideas the receptionist chimed in and shared her personal favorite response: increase awesomeness.

Brilliant! The perfect balance of aspirationalism and measurability. However, Ajax and I agreed — I'm already pretty awesome. So I decided to go with "Enhanced Femininity" instead.

As I giggle-jotted my final answer down, I realized — uncovering epics might not be as easy as I make it out to be in training. I imagine customers probably feel a little like I did when first they come face to face with the question:

"WHAT DO YOU WANT TO GET OUT OF THE SOLUTION?"

But at the end of the day, if you ask 'why' enough times, the answer will yield one of three themes. People are personally motivated by either fame, fortune, or love; businesses ultimately seek to either increase profits, minimize loss, or achieve compliance — a.k.a. stay out of jail. You know you've reached your goal — identified a prospect's epic — when one final 'why' yields one of these thematic responses.

As you work with your prospects to tease out their true

motivations remember: not everyone wants to play the 'why game.' Many people seldom bother to explore motivation at such a high level. Some struggle to appreciate strategy, while others falsely believe overcoming challenges *is* the end game. Whatever the cause, the solution remains the same: go slow and offer leading suggestions.

Suggest that they might choose to use the enhanced visibility an application provides to shorten their sales cycle, or perhaps the improved data integrity will lead to heightened operational efficiency. Once you get the creative juices flowing, your prospect will be quick to share their true motivation, and you'll secure all the ammunition necessary to command a close.

So next time, identify a personal epic. Take a moment to retroactively uncover the motivation behind something you did or purchased recently. Set aside the features you thought you needed; look past the challenges you sought to overcome; reveal your true motivation. Your enhanced appreciation for the process will do more than just make you an empathetic salesman, you'll become an effective one too.

As Seen On TV

I grew up watching TV — a lot of TV — and realized pretty early on that just because a show didn't explicitly instruct you to "[not] try this at home," it usually went without saying. Most of the information you get from television is pseudo-science at best and seldom represents actionable advice. But every so often — if you watch through the right filter — you'll uncover nuggets of wisdom.

In the 80's Jim Henson taught me to overcome the liar's paradox[12] in *Labyrinth*; in the 90's Bruce Willis shared the solution to the water jug puzzle[13] in *Die Hard: With a Vengeance*; and Steven Seagal demonstrated a throw[14] in *Under Siege 2: Dark Territory* so simple, even the comic relief character could do it. Over time my interests shifted from feature films to television series, where the salient lessons are easier to spot; the practicality of advice can often be measured by how frequently it's repeated.

On a slightly unrelated note, in preparation for my Miami migration, I decided to rewatch the first five seasons of Burn Notice. The cutaways of the Miami landscape always make me smile. But this time I got more than future-lifestyle fantasy. This time I managed to assimilate some sage advice which — much to my delight — I got to use soon after.

---

[12] YOU APPROACH TWO DOORS. ONE IS GUARDED BY A TRUTHFUL GUARD, THE OTHER BY A LIAR. ONE DOOR LEADS TO CERTAIN DEATH, THE OTHER PROSPERITY. BY ASKING ONLY ONE QUESTION, TO ONLY ONE GUARD, YOU MUST DECIDE WHICH DOOR TO CHOOSE. WHAT DO YOU ASK AND TO WHOM?

[13] YOU HAVE TWO JUGS — SIZED FIVE AND THREE LITERS RESPECTIVELY. NEITHER HAVE ANY MARKINGS ON THEM. USING AN UNLIMITED SUPPLY OF WATER, MEASURE PRECISELY FOUR LITERS.

[14] I'M LEAVING THIS ONE TO YOU TO GOOGLE.

The Friday after I landed in Miami Beach I called Atlantic Broadband requesting they come install some Internet.

THEM: "HOW'S MONDAY BETWEEN NOON AND FOUR?"

ME: "THIS MONDAY?!?"

Having been conditioned by Comcast, I couldn't believe my ears. Their brilliant service, however, put me in bit of a pickle; I didn't bring a router to Miami with me. Enter: Amazon Prime, who promised to promptly provide what I needed. Except now UPS had to (try to) deliver something to a girl who didn't have a doorbell on a holiday, when the condo office lady — also known as my backup package signatory — had the day off.

I decided the easiest way to solve this problem was stalking — Burn Notice-style. What can I say, I *really* wanted Internet! After the cable guy finished, I took a quick shower, confirmed the package was still "on the truck," and rushed downstairs to keep a lookout for the brown truck of logistics. After standing on the corner for about five minutes — and waiving off like six confused taxi drivers — I realized I needed a better plan. So I marched across the street to the coffee shop, ordered an Americano, and perched myself on the patio.

Without realizing it I had assimilated years worth of Michael Weston's advice. I choose a vantage point that offered a clear view of every approach angle, in an establishment where excessive loitering would be tolerated, if not encouraged. I even packed a magazine to help me blend in. As I sat there nursing my coffee and pretending to read last month's *mental_floss*, I couldn't get Jeffrey Donovan's voice out of my head:

"PEOPLE THINK SPY WORK IS GLAMOROUS, BUT IT'S REALLY A LOT OF SITTING AROUND WAITING FOR SOMEONE TO OPEN A MAILBOX"

This soliloquy mashup looped around in my head like the world's worst pop song, and I realized that the hours and hours of television effort were worth more than just entertainment; television had prepared me for this very moment! As I chuckled at the absurdity of the situation, I couldn't help but ponder the long term impact of the more passive sales and marketing touches that we make.

For those of us who don't sell 'household' brands, we probably spend some portion of the sale defining who we are and why we're legitimate. If you stay at a company long enough your pipeline may mature to the point where that becomes less of a thing. But why?

Perhaps all those business cards your evangelism team have handed out on planes, trains, and trade shows are finally paying off? Or maybe you landed an advocate with a humbly high Klout score? Or perhaps Anderson Cooper just did a segment on the impact products in your industry have on middle America? But my money's on dumb luck.

Whatever the reason, if it advances your goals – don't fight it. The UPS guy didn't even react to my frantic, giggle driven dash across the street; so I shared that I worked from home and offered to buzz him whenever necessary – a symbiosis sure to eliminate off-day delivery worries.

So next time, embrace the power of passive learning. By bonding with your prospect over the circumstance that – however archaic – deposited them into your pipeline, you'll be better positioned to go for the close.

Selective Storytelling

I boast to "live life for the story." But life usually delivers these sales tales wrapped in a shroud of literal. Thankfully my automatic application of sarcasm and 'should've been' cuts through this cloth quite effectively. Yet, it wasn't until my bowling buddy JT shared with me his version of Saturday evening, that I fully appreciated the reflexive nature of my editing.

You know it was a good night when someone calls you the next day to inquire about the location of their belongings. So I happily paused my Sunday football funday to play a quick game of real-life 'dude, where's my car.' After a few minutes of mandatory mockery, I confessed to having no idea where he parked and let the man go about his day. Then I realized I should probably check on Tom — another friend of mine who, last I saw, was with JT.

I'm so glad I called because Tom clued me into the details of the evening that JT selectively omitted — specifically circumstances surrounding the location of his missing keys, wallet, and cell phone. Evidently he didn't spread his belongings about the city like some sort of happy drunk leprechaun on a quest to seed San Jose with treasure, as he alleged. These articles, it appears, were collectively abandoned, still neatly tucked safely away in the pockets of his — also missing — pants!

Personally I respect a man who manages to make it home safely, on foot, pants free, in the middle of the night, walking down the side of a highway. But a man who manages to do all that and still keeps his wits about himself well enough to spin a plausible story in the morning, that's something special. Well done salesman, well done.

Respect aside, the weekend's events should inspire us all. Don't let the specifics of your product's offering detract from the awesome story you tell about it. Only when we really understand what your customers need to decide — or in this case to find — can we adapt our tale accordingly.

So next time, start with the moral of the story. Decide what point you want the audience to retain, carefully select those features or events that support the goal, omit any areas of weakness that aren't absolutely necessary to further your agenda, and commit. Not only will the resulting pitch be more concise, but by avoiding distractions, you'll expedite the entire engagement.

Condo Shopping

After my company-sponsored California relo got a green light, I flew out with one simple mission in mind: acquire an apartment. While boasting about the nine appointments I had lined up while perched on the couch in Gino's office, Clark — a kindred spirit and colleague — walked in to say hi. Gino jokingly suggested he and I team up; Clark had been apartment hunting unsuccessfully for a few weeks now.

Five shitbox-on-a-stick tours later, I reconsidered the joking aspect of Gino's remark. Serendipitously, Clark shared my disappointment in the square footage to price ratio offered near the office. Surely combining our rent rations could furnish a better place. So we decided to give it a go.

During Saturday's tour-de-Cupertino we entertained unfiltered questions, identified an intersection of domicile requirements, and toyed with the minds of several salesman. Anyone who can endure four hours in a hot car, listening to a two song playlist, while fielding my insanity and *not* exit offended and rageful, might actually be roommate material. So we established a plan for Sunday, focused our search, and prepared to do it all again.

I started the day solo, charmed the hell out of an agent named Patti, and secured a verbal offer on what would later become our backup pad. The next several showings only made the morning house seem better. I was about ready to retire the Suzuki's shitty soundtrack, make Patti a happy lady, and hit the pool. But Clark wisely campaigned for diligence, so we refueled and met Fernando at a townhouse off El Camino.

Jackpot! Not only did the unit feature ample closet space and grown-up sized rooms, it had a layout that ensured seclusion on the nights he'd had enough of my neuroses. We made one final stop at an open house to confirm the selection before returning to the office to mooch wifi and fill out the application.

Half way through the paperwork I realized that we should probably make sure that Clark's foxy girlfriend had signed off on this because the last thing I need is to come back from a business trip to a domestic dispute. His casual, "oh yea, she met you at the Christmas party, no worries," response cleared the road ahead; our plan was a go. So there I was thinking how amazingly rational and practical everyone was being, when I get a text:

"GIRLFRIEND VETOED IT."

Naturally I immediately compiled, what amounted to, a pretty strong argument regarding how good of an idea this really was, and went to bed thinking I'd just tell Clark how to sell it tomorrow. Unfortunately he had the same plan, and it didn't work. So come Monday morning, instead of a roommate and a flat, I got Gino laughing, pointing, and singing "I told you so!" like an intolerable toddler.

Oh well, back to square one.

So next time, identify ALL the decision makers early and confirm often. If you, like me, let the deal progress too far without all relevant parties on board, the hole you find yourself in will be way too deep to dig out of. Lesson learned; dicker with deciders or doom your deal.

Exceptionally Eccentric

While cruising along at 38,000 feet one evening and watching *Jobs* – for the third time – I partook in a very fruitful conversation with American Airlines. I wanted to confirm my connection would indeed be on the same plane as this, delayed, outbound flight. Not only did the folks on the American Airlines Twitter desk promise to research the issue, they graciously facilitated a contingency plan.

By the time I finished my meal, all was well flightwise; and by the time the movie concluded, I was well on my way to a Bacardi fueled bout with existentialism. *Jobs* ends with the quote:

> "HERE'S TO THE CRAZY ONES, THE MISFITS, THE REBELS, THE TROUBLEMAKERS, THE ROUND PEGS IN THE SQUARE HOLES... THE ONES WHO SEE THINGS DIFFERENTLY -- THEY'RE NOT FOND OF RULES... YOU CAN QUOTE THEM, DISAGREE WITH THEM, GLORIFY OR VILIFY THEM, BUT THE ONLY THING YOU CAN'T DO IS IGNORE THEM BECAUSE THEY CHANGE THINGS... THEY PUSH THE HUMAN RACE FORWARD, AND WHILE SOME MAY SEE THEM AS THE CRAZY ONES, WE SEE GENIUS, BECAUSE THE ONES WHO ARE CRAZY ENOUGH TO THINK THAT THEY CAN CHANGE THE WORLD, ARE THE ONES WHO DO."

STEVE JOBS[15]
US COMPUTER ENGINEER & INDUSTRIALIST (1955 – 2011)

I make no secret of my fundamental and principled hatred of Apple Computers, but I can't help but relate to this specific diatribe. Many people revere Steve Jobs; they speak fondly of what he's done; they loyally adopt his products; they proudly brag to be a part of his era. The issue I take is that,

---

[15] HTTP://WWW.QUOTATIONSPAGE.COM/QUOTES/STEVE_JOBS/

these are often the same people who frequently tell me to sit down, shut up, and get back in the box. What happened to expecting greater? To saying: "to hell with the status quo"? To denting the Universe?

Look, I'm not suggesting I'm the next Steve Jobs. Far from it. Even armed with a fleeting fit of genius, I'm way too much of a math major — longing for predictable paychecks — to ever start a company. But I'd stand to wager, even as salesmen (or perhaps especially as salesmen) we yearn for a certain consistency in our cash flow and — whenever possible — bank-breaking bonuses.

Perhaps that's why we play the startup game. We idolize innovation and spend much of our time immersed it. So, why do we still find the 'crazy ones' so frustrating? Are our efforts to tame the troublemakers and tendency to repel the rebels, simply our way of masking our own fear of change?

Creativity comes in many forms. Some people even say Silicon Valley was founded by proud residents of the autism spectrum. Having spent most of my adult life navigating the minefield of idiosyncratic behavior many uber geeks lay, I'd believe it. The mind-space provided by an engineer's charm makes way for exceptional development power which yields truly valuable solutions. While there may be little room for manners in the mind of misfits, remember: you hired your engineering sidekick for their brain — deep down you knew that package came with a good bit of geeky charm.

So next time, choose profits over politics. When you loosen the leash, embrace your quirky comrades, and put personalities aside, you'll close more deals, drive more change, and — most importantly — make more money.

Midnight Supplies

My old employer turned ten the other day and their birthday party provided the perfect excuse to pop off to San Francisco for a weekend. In doing so I also gained the opportunity to stop by my favorite dice store. Will – my date for the evening – graciously accompanied me on my mission to make 1-4-24[16] a bartime staple. We arrive at Jeffrey's Toys and Comics where I ask the shopkeeper to bring me the bins of dice. "Sets of six," I say to my confused comrade who still isn't sure why we're there.

I have pretty specific, albeit slightly neurotic, criteria for certain commodities. Suitcases, di, desk chairs, cocktails – all subject to well defined parameters; a good di is pretty easy to define: sharp, not curvy, edges and high contrast numbers. Proper dice, therefore, are easy to read – in even the most dimly lit establishments, and won't roll off the bar all wild-like – thereby forcing you to your knees under some stool attempting to drunkenly retrieve your toy.

As it pertains to the set as a whole, I'm a purist. All six should match, if for no other reason that I'm liable to overthink the fairness of each di otherwise. You know, as if the rolls aren't representing independent random variables. So – sets of six, similarly colored, sharp-cornered dice. Simple enough, right?

Ha!

I don't know if it was my jet lag, Will's hangover, or the fact

---

[16] ALSO KNOWN AS "MIDNIGHT" HTTP://EN.WIKIPEDIA.ORG/WIKI/MIDNIGHT_(GAME)

that it seemed this store stocked dice in sets of five, but it took us nearly twenty minutes to complete this mission. All the while I kept thinking about how frustrating this would be for someone with more particular criteria than my own. The more I dug, the more I had to abandon one color I liked for another; the more sets of four or five I laid out on the counter, the less satisfied I became with any of them. I mean, they're dice. I get that. Who really cares what color they are? But what if I was buying software instead?

When prospects appear with premeditated programmatic expectations, are they doomed for disappointment? Or do problems simply arise when they're presented with too many choices? How can we stop customers from getting five-sixths of the way down a path before realizing they're in a technological cul de sac?

Simple. With guided prospecting and good discovery.

Had the lady asked me what the hell I was up to, perhaps she could have started me with the bin of satisfying dice. Too late now, I sorted it — but think about your prospects. Are any of them wandering down the trial trail without a map? It might be time to offer some guidance.

So next time, lend a hand. As much fun as it might be to watch a blonde with a big grin shake a bin, if you're keen to sell her something — step in before the smiles turn to sighs. By helping customers identify the shortest path to success, you'll see your average time to sale slide in your favor.

Meatball Mission

Paul throws the best Halloween parties. Seriously, the best. And not just because at them, Paul's mom serves the most delicious crockpot meatballs I've ever had the pleasure of tasting. After many years of telling myself I had to score the recipe, I finally made my approach. Paul's mom was happy to oblige. The ingredient list is remarkably simple, only three items long. The first two − making up the sauce − are easy to come by. As was the third: Sam's Club frozen three-quarter ounce Italian meatballs − that is until I moved to Miami and faced a shortage of Sam's Clubs.

I didn't realize this wrinkle would be such an issue until after I promised Ralph meatballs as Sunday's football food one week. You see, Ralph and I have been using Sundays to march our way down culinary memory lane. Previous stops included Captain Nemos-style subs, homemade pulled pork with Neely's cole slaw, my mom's taco dip, and Ajax and Perl's spring rolls. I wasn't about to let something as silly as meatball procurement break this streak.

The search started at Publix where the frozen 'meat' section contained fish. Only fish. I then moved on to Presidente followed by Sabor Tropical where − not surprisingly − frozen appetizers of the Italian persuasion are not stocked in great variety. Winn-Dixie was looking like my last resort, but since I wasn't feeling good about it, I began considering alternate strategies.

Miami must have Sam's Club equivalents, but I wasn't about to join a club just to get one thing. Plus we have zero room in the pantry to accommodate all the other bulk items I'd likely accidentally come home with. I turned to the bowling team for help. Surely one of the guys has a membership to BJ's or Costco or something.

Eddie's wife, Daphne, after hearing the whole story, agreed to help me out, but not before suggesting I try Walmart. Since they are the same company as Sam's Club, it would stand to reason they'd have a similar selection. Friday evening, on my way home from subbing in a different bowling league, I meandered through Walmart and successfully procured some meat.

As I waited in line to checkout, I couldn't help but think about how much the Internet had failed to aid me in this quest. I successfully researched the target items on the Sam's Club site, but that's where website assistance stopped. None of the other stores offered access to a [useful]

digital catalog; they forced me to physically go from store to store. It reminded me of software companies that make you call to ask for more information.

With more and more companies removing the barriers to purchase — letting customers transact online — how do we, as salesman, keep our jobs? Honestly, I don't think we're going anywhere anytime soon, so the bigger question becomes: when your product's value is too subtle to convey solely via a website, how do you get them to call without making them feel forced to shop like it's 1999?

The key is in the content. The more data your marketing team collects as they rank and score your prospects, the more ammunition you have to make a differentiated mark during that first call. When speaking to a human if you don't begin by confirming countless pieces of previously volunteered information, the customer is way more likely to remain engaged. Even though Sam's Club wasn't the ultimate source of my meat, it remains my favorite warehouse club because it strikes the best balance between virtual and brick and mortar browsing.

So next time, write your own recipe for success. When you, as a salesman, add value beyond what a customer can get online or with the competition, you better the odds they will stick with you in the future.

Bustin' Hos

This tale is one of both victory and defeat. A contrast in which, I believe, hides a lesson for us all.

Once again I had the opportunity to 'party' with the Coterie. They were on another mission to de-ho-ify the streets of Miami Beach and — as is standard practice — planned to break the evening into two phases: call-ins and walk-abouts. During the dial-a-date portion of the evening, Wellman escorted an escort back to his room. Things carried on as they usually do, with both parties dancing the line between frank and fiction. But this time extenuating circumstances forced the engagement into an unexpected turn.

Friends from the feds were on-site to assist. This unfamiliar audience caused Wellman to hesitate when asked to demonstrate his commitment to the tale by dropping his drawers. Not because the request was out of bounds, or even outside the scope of standard practices. He hesitated simply because he reflexively considered what the onlookers would think. Through the window of his wonder our girl caught a glimpse of her future; she left before committing any crime.

Contrast this to my run at a similar scenario later that evening.

While cruising the streets, Shale, Rem, and I stopped in a bar known to be target rich. Within minutes Shale was approached, but the girl backed off when I returned with a round from the bar; she had pegged us as a pair. I, unlike

Wellman, give very little thought to what others think of me and was determined to bust a ho that night. So I rolled with it; I flirted a little, flashed a smile, and let her finish her lap about the bar. After a reasonable amount of time passed, Shale reengaged and began collecting the crumbs of the crime.

This wasn't her first sale by any means, and like any experienced sales woman she made a point to trust, but verify. Luckily for me, I make it my mission to always carry evidence to support my stories. Busting out an old Illinois ID that supported our tourist tale left our lady comfortable enough to cooperate in the conversation. She mapped out the details of the evening's offering with sufficient specificity to land her in cuffs – and not in the happy fun way.

Now while the night did have many more busts than bails, the differences between the two outcomes got me thinking about the unintended impact of managerial monitoring. We've all been there – where someone we work for, or simply respect, watches quietly as we go about our daily work. Something about the additional audience makes us feel silly and often screw up things we've done hundreds of times before. I mean to this day, I can't back a car down the street with witnesses watching.

So – in the face of required review, and in the spirit of training the team – how can outside observers abstain from passive interference? And what can we do to better ignore their illusory influence?

Personally I don't blame the fly; it's not their fault they're stuck on the wall. Plus, at the end of the day, prospects – like prostitutes – perceive hesitation as risk. Audience or no audience, we simply can't afford to pause. Remember, onlookers will judge you far more harshly for having nothing to say than if you say something stupid but still manage to keep the deal moving. When you know your product, understand your role, and keenly pursue the requirements — the audience no longer matters.

So next time, stay focused. If you can't think on your feet, take some time to play out several possible conversation paths before you engage. Prepare rebuts, practice plausible alternatives, and — most importantly — with your sights set on the destination, keep pushing forward. If we can learn nothing else from the evening it's that when you commit, you close.

Don't Demo!

So I'm in New York City at the CRM Evolution show a while back and this guy comes into the booth and starts talking to one of the other girls. After about ten minutes, she comes over to me indicating that the gentleman had requested to see a demo and asked if I would mind taking this one.

ME: "SURE! WHAT DOES HE WANT TO SEE?"

HER: "I DON'T KNOW, JUST A BASIC DEMO... HE JUST WANTS TO SEE WHAT THE APPLICATION DOES."

Surely a combination of trade show fatigue and jet lag contributed to her lackluster response. Unfortunately my follow up inquiries regarding what the gentleman does and what his business hoped to achieve with CRM also went unfulfilled. This was the second time in an hour my expectations of a qualified handoff were met with an eye roll. But I agreed to jump in anyway because her lack of inquisitiveness reminded me of the fact that I'd recently been challenged by a few inside sales reps to prove that — what they affectionately refer to as — "Erin's line of questioning" could indeed produce more fruit than frustration.

"WHERE BETTER TO PROVE MY THEORY THAN A TRADE SHOW FLOOR!?!" I THOUGHT.

The three of us walk over to the demo station, make introductions, and my comrade takes off.

ME: "I HEAR YOU WANT A DEMO."

HIM[17]: "YES, I'D LIKE TO SEE WHAT SACCHARIN IS ALL ABOUT."

ME: "OKAY GREAT, BUT I CAN'T JUST JUMP INTO A DEMO; IT WON'T MAKE SENSE OR BE TERRIBLY IMPRESSIVE UNLESS I HAVE A BETTER IDEA OF WHAT YOU ARE LOOKING TO ACCOMPLISH WITH CRM... MAY I ASK WHAT BRINGS YOU TO THE SHOW?"

He smirked shared that he inherited a Siebel implementation that wasn't jiving well with the sales team; his IT organization was directed to fix it and he was debating whether to pull a 'rip and replace' or invest in a repair job. I asked what got Saccharin on his radar, and he went on to share how the morning keynote panel sparked an interest in social.

HIM: "ALL THESE BIG SHOT NAMES IN CRM WERE WAXING ON ABOUT THE MERITS OF SOCIAL CRM, BUT WHEN I WENT ONTO TWITTER ONLY JERRY (SACCHARIN'S CEO) MAINTAINED A RELEVANT, ENGAGED, AND USEFUL PRESENCE."

ME: "DOES YOUR COMPANY HAVE A SOCIAL STRATEGY?"

HIM: "THEY DON'T [YET] BUT I RESPECTED JERRY'S ABSENCE OF HYPOCRISY, AND DECIDED TO STOP BY."

Further questioning led me into the inevitable discovery loop, where the prospector becomes trapped in the logic that simply having a better, more unified, system *is* the goal. So I invited him to take a step back and explained that whatever CRM he chooses, it will obviously meet his base requirements, and gather all the information his constituency desires. I then encouraged him to recognize that knowing – that is, having data – isn't enough.

---

[17] JUST TO BE CLEAR - 'ERIN'S LINE OF QUESTIONING' **DOES** INCLUDE ASKING THE PROSPECT THEIR NAME, BUT UNFORTUNATELY 'ERIN'S BRAIN OF DISTRACTION' SELDOM HANGS ONTO SUCH INFORMATION. AND SINCE I CAN'T REMEMBER THE BLOKE'S NAME, I'M JUST GOING TO CALL HIM 'HIM' FROM HERE ON OUT.

He should seize the opportunity to discover what KPIs[18] the sales team is measured on — like, perhaps sales cycle duration or close percentage. Because if he knows how they measure their own success, not only will he design future processes more intelligently, he will better prioritize and phase use cases. And most importantly, he can align himself with a measurable improvement and thereby avoid subjective scrutiny regarding success of the project.

At this point I conceded that it was okay he didn't know which KPIs in particular his sales team treasured. He promised to investigate, and we agreed — for the purposes of this demo — to discuss the seemingly universal goal of shortening their sales cycle. I demoed a couple screens, and highlighted how Saccharin's user-first philosophy drives better behavior — which leads to shorter sales cycles than the management-first focus of Siebel.

Because I first dropped a KPI anchor, I successfully deflected some questions regarding specific ERP[19] integrations on the merits that 1) by using APIs[20] you certainly *can*, and 2) the integration he's describing (by his own admission) doesn't have a clear and direct correlation to an expedited sales cycle. Consequently such integration efforts often find their way into phase two initiatives. Which — because he intends to tie phase one to a measurable revenue boost, and not just 'increased efficiency' — will be easy to fund. Together we revealed the key to year over year returns.

This is when he thanked me. Twice.

---

[18] Key Performance Indicators
[19] Enterprise Resource Planning
[20] Application Programming Interfaces - the means by which one software application talks to another.

I believe the quote was:

> "I CAN'T TELL YOU HOW MANY TIMES I'VE ASKED FOR A DEMO AND
> THE GUY JUST SHOWS ME A DASHBOARD WITH A PIPELINE CHART,
> LIKE THEIR CHART IS SOMEHOW MORE PROFOUND THAN THE OTHER
> NINETEEN PIPELINE GRAPHS I'VE SEEN.
>
> THANK YOU FOR TAKING THE TIME TO ASK SMART QUESTIONS. I'M
> DEFINITELY GOING TO INCLUDE SACCHARIN IN THE SELECTION
> PROCESS BECAUSE I SEE NOW THAT SEIBEL ISN'T GOING TO WORK.
> THANKS AGAIN FOR NOT JUST JUMPING INTO THE DEMO, AND TELL
> JERRY I'M FOLLOWING HIM ON TWITTER."

So next time, don't dash to the demo! There's nothing wrong with saying no when you know you're ill-prepared to deliver a compelling demonstration. Your customers will respect you more when you have both the courage and the conviction to assert a solution instead of just another a silly chart.

Read the Signs

I had a follow up appointment with Dr. G. the other day; she wanted one last look to confirm all my ankle bones had healed on schedule. Since my foot felt great — and I knew the visit would certainly confirm that — I decided to ride my bicycle to her office. During my four mile ride to Mount Sinai I kept thinking how nice it is that I'm back to my old routine so quickly. Then, about three miles in, I realized I didn't actually know how to get there.

Now, that's not to say I didn't know where I was going — I could easily point out the hospital on a map; macroscopically I also knew where I was in relation to that point. From a first person view of the island, however, that perspective lacks practical application; Miami Beach's grid system erodes as the island's shape becomes increasingly irregular. Not to worry though, I sort of paid attention all those times the cabs shuttled me up the beach — I'd be fine.

Yea...

You see, those fleeting memories quickly became replaced by Chicagoan instinct. I knew I was approximately five blocks south and four blocks east of my destination when I decided to start zig-zagging toward the hospital. Head little north, then a little west, then north again — I'd just ziggy-zaggy my way there. Great plan, right?

I hit a light at 41st Street, and per Erin's cartesian navigation protocol, I turned west to keep moving. I shaved off all the westerly requirement before I realized I accidentally got on the highway. Not like the time I accidentally left the island on a causeway with a bike lane. This was an actual proper highway. ON A BICYCLE!

I spent the first quarter mile in denial, telling myself the signs were just kidding and this was *totally* the road that the cab got on that one time; surely it had a secret turnoff toward Mount Sinai. As I rode the shoulder I could see my destination through the fence on my right. I longed for a turnoff, a feeder, a service road — something that would let me escape. When I saw the sign that said the next exit was in three miles, I could deny it no more — I was officially biking down the highway.

Time for an exit strategy!

My thought process went something like this:

> "SHIT I CAN'T TURN AROUND, THAT'S MORE DANGEROUS THAN THIS!"

> "THERE'S NO WAY TO GET TO THE OTHER SIDE TO RIDE BACK WITH THE TRAFFIC FLOW BECAUSE OF THE CEMENT DIVIDER, AND THAT'S A SUPER STUPID IDEA ANYWAY."

> "IF I DON'T GET KILLED I'M TOTALLY GOING TO GET ARRESTED!"

> "WHY DO I ALWAYS FLIRT WITH LEGAL IRRESPONSIBILITY ON TUESDAYS[21]!?!"

> "I WONDER IF I CAN GET MY BIKE OVER THAT FENCE."

> "THE SMART THING WOULD BE TO CALL FOR A RESCUE."

> "WHAT'S THAT LIKE A SEVEN FOOTER?"

> "I CAN PROBABLY LIFT THE BIKE THAT FAR."

---

[21] (WHEN NONE OF MY COP BUDDIES ARE WORKING.)

"Plus I can't ask for help, I'm no pussy, and even if I had someone I could call, help would ruin the story."

"Okay let's give the fence a try."

"If I pull this off it's soooo becoming a *Sapient*."

So I turned off the road, breezed across the grass, threw my bag over, leaned my bike up against the fence, and started climbing. Once over, I stood on the other side assessing the situation. Any bicycle recovery maneuver would mandate that I reach – not lean – over; the fence's top didn't look it, but turned out to be quite pointy and sort of sharp.

By now I had an audience; some dude in the hospital parking lot decided to stop and gawk, so just in case he chose to rat me out, I decided to hurry. I climbed back up, reached over, grabbed my handlebars, and temporarily forgot I'm not a ninja. I figured I could do like those Army guys do on TV – if I just fall, my weight will pull the bike over.

Horrible plan.

The handlebars hooked into the chain links which left both me and the bike just hanging there on opposite sides of the fence. I laughed it off, climbed up again, and reached for the handle bars. Plan B time: hulk it. I'll spare you the play by play of how I took the title for single-armed highway-side bike hoists because as my maneuver materialized I realized this is probably how sales guys get their forecasts so turned around. They fail to read (or accept) the signs!

How do we avoid cruising along a sales cycle selectively oblivious to the signals our customers give? Or perhaps the problem isn't that we don't read them. Maybe we're just unwilling to accept the truth when it's staring us in the face. What makes us believe the obvious doesn't apply to us? Whatever the cause of our naivete — be it selective or circumstantial — no good comes from pedaling forward underinformed.

So next time, pay attention! Autopilot might be great, classical approaches may feel proven, and instincts may have brought you this far — but today's a new day. When you focus on what's actually happening around you instead of daydreaming your way about the world, you won't end up like me — thirty yards from the finish with no good path to pursue it.

Measuring Value

I was just sitting around enjoying the sunshine when a buddy of mine called to tell me he broke it off with his lady. The news surprised me because last I heard they were having a great time and he seemed genuinely happy.

> **ME:** "WHAT HAPPENED TO 'HER COMPANY IS CAPTIVATING' AND 'SHE SHAGS SUPERBLY'?"
>
> **HIM:** "IT IS, AND SHE DOES, BUT I'M NOT GONNA MARRY HER."
>
> **ME:** "BECAUSE ..."
>
> **HIM:** "SHE'S YOUNG, NOT READY TO COMMIT YET. AND YOU KNOW ... CLOCK'S-A-TICKIN'."
>
> **ME:** "DUDE – SHE ROOTS FOR YOUR FOOTBALL TEAM. WHO CARES IF SHE'S TWENTY-FIVE AND DOESN'T WANNA GET MARRIED THIS WEEK. WHY NOT JUST ENJOY YOURSELF?"

See I'm of the opinion that – with personal relationships in particular – the investment *is* the return. I mean when you make a new friend, do you immediately think to yourself:

> "WELL, IN TEN YEARS THIS 'FRIEND' IS PROBABLY GOING TO MOVE TO THE BURBS, BUY A CAT, AND USE DANDER TO PREVENT ME FROM VISITING... SO WHY BOTHER JOINING HIS BOWLING TEAM? THE FRIENDSHIP CAN'T GO ANYWHERE; I'D JUST BE WASTING MY TIME!"

Of course not!

Despite concurring with my assessment of the friend scenario, this guy insisted that his love life demanded different standards, and it made me wonder if certain technology purchases bear comparable capricious criteria.

Do prospects realize not all return is quantifiable by a system's report? How can we measure and articulate the value of live-action gains that don't obviously accrue? What makes a product, person, or project worth pursuing?

Like a recently divorced, middle-aged man, a customer moving off antiquated technology is more inclined to concern himself with the now. They've seen firsthand the year over year impact of being contractually bound to a tool that no longer suits their needs.

Whereas technologists, who've enjoyed the freedom to bounce from solution to solution, may believe their experience has left them ready, able, and required to make a longer term investment. These guys are harder to sell to, for they are a tricky mix of cocky and over committed − believing themselves wise enough to anticipate market maturation, while remaining so eager for recognition they demand unrealistic returns.

If your company journeyed all the way to the proverbial chasm by rescuing the recently released and now face selling to people who − like my buddy − are so focused on the destination they can't see the road, please refer to the timeless advice of Ferris Bueller. Because, technology, like life, moves pretty fast... if you don't stop and look around once and a while, you'll miss the return.

So next time, recognize incremental value. It takes a lot of good days to build a successful romantic partnership; it takes many small changes and marginal improvements to accrue big returns. Help your clients understand the value in today and they'll surely understand the value that will come with time. Win today and you'll win tomorrow.

The Catalog Conundrum

While thumbing through *SkyMall,* I discovered a product described as "the world's first cordless window vacuum." Unfortunately the description doesn't make clear whether the revolution laid in the fact that this manufacturer removed the cord, or whether their genius simply applied cordless vacuum technology to windows. Several additional questions flooded to mind. How much suction does it take to remove the streaks that knock-off Windex leaves behind? Would this work on tile too? Or would the grout nooks undermine the suction?

Despite representing a great source of amusement for me, I don't often buy things from catalogs. Probably because I'm so easily distracted by the constant sardonic commentary coursing through my head, and the medium lacks the conversation skills necessary to present a reasonable rebuttal. But I wonder, when we let our prospects shop unsupervised, do we run the risk of their cynical sense of humor coming between them and a novel solution?

I'm reminded of a classic *Seinfeld* bit where Jerry, while trying to figure out if he's invited to a party, asks:

"HOW DID HE SAY IT? ..."

"WHY WOULD *JERRY* BRING ANYTHING?"

-OR-

"WHY WOULD JERRY *BRING* ANYTHING?"

Especially through written communications, it's difficult to clearly discern the intended tone of a message. With ambiguously worded assertions in particular — where emphasis on one word versus another can completely alter the meaning — we often leave prospects with nothing more to go on than their own internal voice. It kind of makes you wonder whether or not your sales-mails sing on key.

Too often we take marketing for granted; different materials are designed to target different audiences, using a carefully calibrated tone and tempo. As such, we as salespeople, must use caution when choosing which message, which materials, and which value proposition to share with a prospect. For if we don't, we risk them developing a suboptimal opinion.

You might be surprised to realize, the famous demo faux pas of 'show up and throw up' can rear its ugly head via email as well. When you send everything without regard for your audience — just for the sake of 'staying in touch' — you're just as bad as that sales engineer who lacks a pause button when he demos. Thankfully, the quest to avoid this bad practice is — yet again — fueled by proper discovery.

When you know what they need to know, and you understand the messages contained in your myriad of materials, you'll be well prepared to choose only the pieces that make a clear, consistent point. Knowing, for example, whether your audience longs to clean windows without the burden of a cord, or rather free from squeegees is critical — both to message and to method.

So next time, share succinctly. Unveiling only the relevant truths with a thoughtful tone will keep your prospects focused and ultimately shorten the sales cycle.

Booster Shot

Holiday support coverage means working odd hours, but it's not all bad. I decided to volunteer for the late shift because I knew it'd free up my morning. Ordinarily I'd just sleep till noon before rolling out of bed to get on a call, but this time I was determined to generate some positive juju. So I decided – in the spirit of 'operation experience daylight' – to make an appointment to get my hair cut.

That kept me busy to about 10:15.

Then, so as not to immediately morph back into a homebody, I began inventing errands to run. I started by returning some crap to Menards that I had in my car for about a month. I don't much care for Menards; it's a very weird store with very old crap – even the nails are rusty. The proximity to the oxidized equipment reminded me of the last time I got a tetanus shot, which quickly snowballed into a quest to get the booster I needed like five years prior.

Upon arrival at the Take Care Clinic, I learned that Illinois recently had an outbreak of Whooping Cough! After resolving to avert similar surprises by starting to watch local news, I obliged the nurse and took the vaccine that included Pertussis. The problem, however, is the whole thing made me wonder what other vaccines I might be missing out on. When I got home I went on the CDC's website and – in a fit of hypochondria – looked up every country I've visited in the past year to see what else I should have been immunized against. I tell you what, the list is not short.

Unfortunately it turns out I already pretty much exhausted Walgreens' catalog, save one: Hepatitis A, which – by the way – is recommended for pretty much anyone who travels anywhere fun. So I did what any impulsive, paranoid nerd would do; I went back to the clinic for round two.

On the drive over however, I began to worry they wouldn't sell me the second one. I worried they might mistaken me for some sort of vaccine junky. Which made me wonder if this happens in sales too? How many opportunities do we lose because our customers are simply too embarrassed to admit they forgot to ask a question or buy enough on the first try?

We've all been there, you walk out of the grocery story only to realize you forgot the one thing you went for but decide you don't care enough to bother going back in. The store is at a loss here because they don't even know they've missed an opportunity for a sale. Suddenly the subtle genius behind the "did you find everything okay" becomes quite clear. Because at the end the day, we are in business to make money, not judge our customers.

If I hadn't walked in looking guilty still wearing the bandage over the original injection site, there's a good chance the woman wouldn't have even noticed I was the same person. The nurse simply indulged my request and didn't even make me regale her with the justification story I cooked up on the drive over.

So next time, give your customers an out. Your prospects likely already feel bad for missing out on any expired promotions. Skip the guilt trip, divert the judgement eyes, hold back the snicker, and just take their order.

The Voucher

At about 18:50 on an otherwise ordinary Saturday night I cleared security at SJC[22]; as I made my way to the gate I heard an announcement soliciting volunteers to surrender their seat on my flight. Without breaking stride I strolled up to the counter and told the agent I wasn't in a hurry to get to Orlando. The other agent gleefully squealed a little and said:

"ORLANDO? GREAT! WE CAN GET YOU ON JETBLUE."

I immediately became very concerned with what that meant for my upgrade; it's a long way from California to Florida and fucked if I was going to spend it in coach. So I asked if I could just take the United flight to LAX and continue on as planned from there. He said "you can ask," so I told them to watch my bag while I went to check.

During the stroll (because I really do hate flying United) I pulled up my trusty American Airlines app to search for other potential paths to sunny Florida. When the conversation with the United counter went about as well as I expected, I set my sights on AA 272 which left SFO for MIA at 20:40 and ultimately got me to Orlando at 08:00 Sunday morning — just in time to bum a ride in a colleague's rental car. "Perfect," I thought as I relayed this information to the agent. I then confirmed they'd pay to schlep me the thirty-five miles from SJC to SFO and reminded him to add me to the upgrade list.

After a few moments of classic airport-counter-style frantic

---

[22] Mineta San Jose International Airport in San Jose, California, USA

typing, we had a deal. But there was a problem. While forty-six people were booked on the flight, as of 19:10 only forty-four had checked in, so the system wouldn't let him issue the voucher yet.

"No worries," I said, "just let me know if you need me."

Ten minutes later I decided to get on the plane, and reminded the team if anything changed they could find me in 1A.

At 19:28 they did!

I gathered my things and scurried back up the jet bridge, confirmed my continued interest in the Miami route, and started the paperwork. With a travel voucher in one hand and a cab coupon in the other, I headed for the curbside pickup point. We departed SJC at 19:58; the race was on! For those of you not keeping score, at this point I had effectively taken a $400 bribe in exchange for the opportunity to switch airports in a under forty-two minutes.

Naturally I tweeted about the impending, well-funded adventure. @AmericanAir quickly engaged and offered assistance. While I can't confirm any communication was actually relayed to the gate agent at SFO, it felt good to know someone with contacts on the inside was listening.

With traffic karma on my side, I flew up the 101 expressway, cut in line at TSA, placed first in the 100 yard flip-flop dash down terminal two, and arrived at the gate to discover my upgrade − delightfully − came through! The night ended better than expected; I managed to squeeze in a pseudo workout, score dinner from one of the friendliest flight crews I've had in a long time, and even parlayed the story into an invite to the Admirals Club as my seat-neighbor's guest.

I arrived in Orlando the next morning freshly showered and properly caffeinated. As I made my way to fetch my suitcase from its solo jaunt via LAX, I couldn't help think about the incredible team effort that went into this successful journey. Then I realized, in most deals, there are more parties of influence in play than we might appreciate.

As much as I'm always on the lookout for an adventure, had I not trusted American Airlines to reunite me with my bag, or reroute me [again] should traffic have not cooperated, I wouldn't have agreed to give it a go. If your prospects don't trust you, your team, or your company, they aren't likely to jump at opportunities either.

So next time, make sure your prospects feel supported. Remind them that behind you stands, not only a solid product, but a team of dedicated agents, trained twitterati, and agile account managers − all eager to ensure their success. Where there is trust, there are contracts.

Differences

Over dinner Ralph told me about an invite we received to go to a friend's BBQ the following day. Apparently Kevin, the host, had just texted to confirm our noon arrival time. Ralph said he let him know we couldn't do noon because of our plans to go race stock cars and attend a gun show in the morning, but that would be happy to come by around two in the afternoon. Kevin agreed to the revised timeline and our Sunday was officially planned.

The next day, while waiting for the classroom portion of the racecar experience to commence, I asked Ralph if Kevin had anything to say in response to his text.

RALPH: "HE SAID 'COOL.' I TOLD YOU THAT YESTERDAY ..."

ME: "YEA, WELL... I DON'T KNOW, I THOUGHT MAYBE HE HAD MORE OF A REACTION OR SOMETHING."

RALPH: "WHAT DO YOU EXPECT? HIM TO TEXT ME BACK: 'WHOA, GIMMIE A CALL BRO, WE NEED TO DISCUSS THIS!' ..."

The thought of such a response made me laugh out loud. For a while. Once I collected myself, we went on to discuss how texting has severely crippled the follow up question market. Conversations are much more matter-of-fact when practiced 160 characters at a time. While one could argue that our differing expectations are more gender related than format driven, I couldn't help but think about how often clients 'tee one up' for us and we neglect to swing at it.

I don't know about you, but I think driving a racecar is pretty damn cool; if I told someone about it, I'd expect it to generate at least a little bit of curious interest. Similarly, when a customer says something like "my clients aren't returning my calls" they probably expect us to respond with something more consultative than "that sucks." So how do we know which dangling remarks are remotely relevant and which are really just rants?

You just need to ask. Sometimes a simple "oh yea, tell me more..." is all it takes to uncover discovery gold. Sure, Kevin's brevity ultimately worked in our favor; by the time we arrived the story had matured into something far more interesting than mere inertial intentions. But when you're selling software, selling solutions – it's the future plans that matter. Without first knowing how your clients want things to play out, you can't paint a proper vision of the future for them.

So next time, take a swing. When your prospects tee up a story, ask them to tell it. A few strategically placed 'whys' and you'll be well on your way to a solid solution story.

Productivity

I get a lot more work done while sitting on the balcony with my Chromebook than any other [quasi] indoor location. I'm not sure if it's the scenery, the fact that I have no instant messenger service out there, the relentless freezing of ABC's iPad app, or — as an episode of *Better off Ted* suggests — the fidgeting induced by the uncomfortable chair. Whatever the reason I manage to plow through projects at a much better clip on the balcony than I ever manage at my actual desk.

So I guess it's a good thing I live in Miami now, 'cause I tell you what — if I still lived in Chicago, where the sunshine runs out long before the assignments, and heat is hard to come by — I'd be back to being the apathetic stressball of yesteryear. To tell you the truth, I didn't realize how much I missed these little fits of productivity until now, having reunited with them.

Back when I worked at Saccharin I had the autonomy and passion to create things. I used to write stories, record videos, create presentations, and experiment! But as they grew up and I moved on, I realized jobs in the 'real world' are horrible. Horrible. People want everyone to stay in a box so badly that bosses frequently ban me from inventing. Somehow it became better to do nothing than to do anything other than a lame, cookie-cutter demo.

A rising tide lifts all boats? More like: don't rock the boat; it makes me look bad. And it made me wonder, when a business decides they need to breathe new life into their systems and processes, is it because the technology on hand is stale? Or has their team simply become so disenfranchised that they've sucked the life out of existing solutions? As salespeople, pitching products that change processes, how do we differentiate between people problems and software shortcomings?

In reality this is probably paramount to a defense attorney conveniently neglecting to care about a client's guilt. You sell software, so if it's simply a people problem — acknowledging that fact could kill the deal. But if you sell something as a service, ignoring this fact will cost your firm more in the long run. Which begs the question: is it even worth the time and risk of trying to find out, or is there another — a better — way?

Instead for digging for answers you're not sure you want or need, work on your spin game. Understand that new technology can drive changes in your people *and* their processes. In fact, that's the whole point. Help your clients understand that both lackadaisical laborers *and* lackluster software hinder progress.

So next time, promote fits of productivity. Help the team — first yours, then your clients' — feel empowered. By providing the tools necessary to support these new found fits, you will deliver the breath of fresh air necessary for everyone to move forward, stop suffocating, and finally produce everything they promised.

Escape Artist

After driving along the rental return road for so long that I actually thought I had missed the turn off, I finally arrived at the Hertz counter at DFW[23]. Happily, I handed off my borrowed Fiat. While waiting for the lady to issue me a receipt I decided to get ready for pre-check. So I chugged what was left of my water, retrieved my driver's license from my wallet, and placed it in my phone holster for easy access. Then, with my receipt confirmed and secure in my inbox, I put my phone away and boarded a bus headed to Terminal C.

Just as I sat down I realized I didn't know if there's a TSA Pre checkpoint in C. To address this question, I turned to the Internet, but when I reached toward the holster to whip out my phone and begin to Google, my phone pocket felt thin. My phone was secure, but there was no sign of my driver's license!

I looked in my pocket, under my phone, under my butt, and under the seat. By the third stoop and twirl I was pretty sure the other passengers thought I was insane. I managed to catch the eye of one of my bus-mates and asked him to confirm there wasn't a Florida driver's license hiding behind me. There wasn't; it was official: somewhere between the Hertz counter and the bus loading zone, my driver's license escaped.

Over the next 30 seconds, the following ran through my head:

---

[23] DALLAS FORT WORTH INTERNATIONAL AIRPORT

" Seriously?!?"

"Man, I don't want to ride this stupid bus back."

"I wonder if I can talk my way through TSA."

"What are the odds of it still being there anyway?"

"Between the gun permit, press pass, credit card, and my vast knowledge of my personal information, they'll *totally* let me on the plane."

"I wonder if we can just radio over there ..."

"But then I have to talk to the bus lady ..."

"I really don't want to talk to the bus lady."

"Wait, I still have my passport with me from when I went to Canada!"

"Fuck it – I can get home!"

After retrieving my passport, I sat back down to plan my next move. Part of me still felt like the responsible thing to do was attempt a rescue mission. I mean I had the time, but that's a lot of work and I really didn't want to be one of those people rushing about the airport like an amateur. So I decided to explore a path of lesser resistance.

The really pathetic part of this whole story is the fact that I had just bragged to Ralph two days prior about how my driver's license had a gold star. By teaching him that a gold star in the State of Florida means you can order replacement IDs online, I effectively jinxed myself. Oh well. I decided if I could order a replacement online before I got to the terminal – I'd abandon my lost license.

Sure enough, even after typing my CVV[24] code wrong on the first try, I beat the bus. What a testament to the simplicity of the Florida DMV's website. As I waited for my order confirmation email, I began to think about how people choose one solution over another. As prospects weigh their short term goals — make my flight, over their long term ones — drive home legally, how can we influence where they perceive balance?

All else being equal, how can we tip the scales in our favor? When selling to line-of-business folks, how can we stop them from abandoning the vision for the sake of simplicity and a quick win? Because at the end of the day people are lazy. Perhaps the answer, then, is to simply make the buying process simple.

So next time, give 'em an easy out. After all the selling is over, before you send them off to simmer, set up their proverbial shopping cart. When completing a purchase, or accomplishing a goal, takes but a few clicks — instead of a pen, a lawyer, and a UPS man — certainly the scale will tip your way.

---

[24] YOU KNOW, THAT THREE DIGIT CODE ON THE BACK OF YOUR CREDIT CARD.

That's not All

"I'm allergic to smells" is probably the most poorly architected sentence I utter, and I easily say it once a month. The sensitivity is severe enough that I actually purchased a year's supply of Fresh Rain scented All laundry detergent — the only flavor that doesn't actively make me sneeze — from Walmart when I moved to Miami. Please, don't mistaken the stockpile as an audition for an episode of *Extreme Couponing* or anything; it's just not an easy flavor to find.

So you can imagine my delight when, months later, I started seeing this seemingly obscure scent at stores like Publix and Target. I picked more up, just in case; I couldn't pass up the opportunity to cut Walmart — the store second only to Apple on my retailer hate-list — out of my life for good.

Flash forward a few weeks, I go to do laundry and I swear, All changed the smell. The smell I've relied on for YEARS. Not ready to accept this fact, I spent most of the night searching for the odor's 'real' source around my apartment. I kept telling myself it couldn't possibly be the detergent. All wouldn't do that to me. It goes against the brand. It undermines my loyalty. It means I own about 132 loads worth of soap I can't use!

::sigh::

I decided to call it a night and resume laundering in the morning, hoping the fragrance would prove phantom. The next day, while waiting for the spin cycle to terminate, I caught myself reading the bottle:

"With improved scent."

Lies! First of all, who are you to tell me what's improved? Secondly, if you're going to muck with the fragrance, change the bloody name! Would Fresh-er Rain have been *that* hard to print? Anything to signal to loyal purists that they were about to purchase a giant bottle of nose tickle!

The whole mess got me thinking about brand purity. As companies grow, sometimes rapidly, strain is placed on both the culture and the product. Many growing startups talk about 'preserving their culture' and tout how they 'honor their roots' – but what about us old timers? How do we communicate a new marketing message to our existing customers when we barely have the schtick down ourselves? Moreover, how do we explain dramatic shifts in product or roadmap to those who've staked their reputation on our continued supply of a predictable product?

Personally I think it comes down to transparency. Had the Sun Products Corporation not so grossly mishandled the change, I might not have been so eager to take Ralph up on his suggestion to explore more 'natural' soaps at Whole Foods. Customers will only forgive small changes when you trust them enough to communicate the changes up front. We all understand that the market shifts, but tick someone off and the likelihood of them exploring alternatives within your product line plummets.

So next time, don't bury the lede. When you rely on fine-print to disseminate 'improvements,' you disrespect the most loyal of your customer base. Remember people are lazy and spending an afternoon sniffing soap, while fun, takes time. Save 'em the bother; be upfront.

Hesitate

An architect, an engineer, and an evangelist walk into a coffee shop and the barista asks:

"ARE YOU WITH FRAMISH?"

The 'sales' engineer blurts a commanding "No." I, still unable to shake the horrible aussie-adjacent accent I acquired during a recent trip down under, contradict him with a drawn out delivery of the word yes. Lock, one of Saccharin's more entertaining software architects, rounded out this clown show of responses by asking the lady to repeat the question.

Luckily the glare that accompanied my retort confused the boys just as much as the shenanigans did the coffee girl. So I used the collective pause to secure spokesperson status. She repeated the inquiry, twice, but I still had no idea what she was asking specifically. Framish? Farm fresh? From which? None of my internalized guesses made any sense.

All I knew was whenever someone who's about to hit total on a cash register stops, looks up with big eyes, and says "wait are you with so and so"... the correct answer is "Yes!" To try to rationalize our group's confusing first attempt, I explained:

"SORRY ..."

"THIS GENTLEMAN (LOCK) IS, YES..."

"IT'S JUST LEE AND I AREN'T, AND HE THOUGHT YOU WERE ASKING ABOUT ALL OF US."

She smiled, said okay, and scurried to get someone else. The next thing we knew the bill got ten percent lighter. I kind of hope the manager went the extra mile and annotated the affiliation on Lock's frequent coffee-er profile for perpetual application. I doubt their CRM is that sophisticated, but we can hope.

So next time, hesitate. When a prospect poses a question, pause and ponder your next move. In sales situations many questions have a clear, predetermined, 'correct' answer and – as a quality salesman – it's on you to sort that out before you respond. In the coffee shop, worst case scenario, we'd just say we left the proving paperwork in the office. In a serious sales situation, however, one wrong answer could cost you more than quarter off a coffee, it could cost you your quota.

When Repetition Gets Risky

I have a way of repeating stories verbatim each time I recount them, and it proved quite handy during my initial transition to sales. To command a pitch with the certainty and expedience required to deliver it before your subject reaches their proverbial floor takes time and practice. Such an elevator endeavor is only complicated by your desire to sound more like a human than a robot. Yet even after you've successfully achieved the perfect balance between content and spin, a new risk arises: you forget to – or simply stop – listening to yourself.

Take something as simple as hearing the specials at a restaurant. After what I'm sure was a painstaking cram session, the waitress for my final Sydney meal marched over and declared that I could enjoy the fish and chips either battered or fried. Anyone who's ever eaten with me knows this kind of announcement is not something I can pass up. The exchange went something like this:

ME: "WAIT, EXCUSE ME, IT'S A CHOICE OF BATTERED -OR- FRIED?"

HER: "YES, WE OFFER THE FISH EITHER DIPPED IN BEER BATTER OR GRILLED."

ME: "OH, SO FRIED OR GRILLED?"

HER: "NO, THEY ARE FISH AND CHIPS, AND WE OFFER THEM BATTERED OR FRIED..."

I loved how she said "no they are fish and chips" like *I'm* the idiot in this conversation. We continued for a few more rounds of me inquiring about how the hell the batter would cling to the fish without frying before maintaining a straight face grew too difficult and I disengaged. Sadly even following the conversation I'm confident the waitress still had no idea what I was going on about, and remained doomed to repeat her inaccurate shtick for many tables to come.

This fine girl clearly missed the lesson buried in my sense of humor – and probably spit in my food to thank me for trying. But we can reflect on the wisdom and ask ourselves: what about my pitch doesn't hold water [anymore]? Have I sang this song so many times the wrong lyrics sound better than the correct ones?

Over time the market, the tool, indeed the world changes, but too often we continue to plow forward with the same old agenda. Remember, your opening statements – whether you qualify leads, manage relationships, or engineer sales – shape the opinion of the consumer in a way that is difficult to undo. Take it from someone who spent ten adult years thinking the saying was "Cadillac reflexes[25]" – every so often, it's good to verify that what you 'know' is as truthful as it sounds in your head.

So next time, update your spiel. A quality pitch incorporates the value described by today's thought leaders and guides your prospects toward a shared appreciation of the subtle complexity of their mission. Create the bonds necessary to complete not only the buyer's journey, but also the continued quest toward making them your most successful reference customer to date.

---

[25] AS OPPOSED TO "CAT-LIKE."

Temporary Solutions

Following another uneventful business trip, I landed at MIA, grabbed a cab home, powered up my computer, hopped in the shower, and got ready to go out. Just like I always do when I get home from the road. But this time, just as I'm about to walk out the door, I think:

"SIGN IN TO WINDOWS REALLY QUICK JUST IN CASE SOMEONE WANTS TO LEAVE YOU AN IM."

So I clickity clack my heels over to the computer and shimmy the mouse. Nothing. I shimmy it again. Nothing. I wait... hit space... wait... still nothing. I look around. The lights are on and I can hear the fans. By now surely it must be booted, I mean it's been twenty minutes.

I turn it off and back on again only to realize it's not beeping. Two hours of tourettes troubleshooting later, with my hands in the air, I pronounced my computer dead. I spent Friday frustrated. All I really wanted to do was plug my hard drive into another machine to confirm its integrity, but I was surrounded by a bunch of freebie iPads and a work machine with a unfriendly USB security policy.

Internet, Internet everywhere and not a port to sync!

"THIS IS WHY YOU NEED TWO COMPUTERS!" I EXCLAIMED WHILE GESTICULATING WILDLY AND SHARING THE STORY WITH SHALE.

That night I faced a choice — rush order a computer that would be delivered after I get home from my upcoming jaunt to Nashville, or ... actually at this point I hadn't yet identified a suitable alternative. It wasn't until the next day when faced with stupidly exorbitant rush fees did I cynically say:

"DUDE I CAN PROBABLY BUY SOME POS[26] FOR LIKE A $150 AND NOT HAVE TO BOTHER WITH THIS RIGHT NOW."

South Florida isn't exactly oozing with computer stores — or über nerds with spare PCs for that matter; my shopping choices boiled down to which TigerDirect location I preferred. So off to Aventura I went.

I marched in and asked the bloke to direct me to the off-lease machines. The signage on the stacks detailed the bundle you "had to" purchase to obtain the machine. The website made no mention of such a scam. Three people came by to see why I was staring at the box stacks with the face of a disappointed toddler. I explained my disgust with the online-offline discrepancy to each of them, all of whom insisted the manager would have to make the call. Yet none were willing to summon the allusive manager for me.

Much to their dismay, ignoring me doesn't work. I spent the idle time pulling up the site so I could insist they honor the advertisement. When finally faced with a managerial audience I explained that I just needed a POS machine to tide me over while I got a proper 'puter put together. He responded by immediately trying to sell me their assembly services:

"I'LL EVEN SELL YOU THIS ONE WITHOUT THE BUNDLE IF YOU USE US TO BUILD YOUR NEW ONE."

I laughed. He didn't. So I asked a flurry of 'think it through,

---

[26] PIECE OF SHIT

yo' questions to help him along:

> "YOU ALREADY CONFIRMED THAT THE HELMET I'M HOLDING MEANS
> I'M ON A MOTORCYCLE ... SO HOW EXACTLY AM I SUPPOSED TO GET A
> FULL-SIZED MACHINE HOME?"

> "DO YOU EVEN HAVE CASES IN THE DIMENSIONS I WANT?"

> "WHY WOULD I NEED THE TEMPORARY COMPUTER, IF I'M GETTING A
> NEW ONE FROM YOU TOMORROW?"

After ten minutes or so, he caved and consented to my computer-only purchase. As one of his flunkies supervised my walk to the cashier, I got criticized for throwing a hundred bucks at a temporary solution to this problem. As if *that* was the yuppiest thing I could possibly do.

> **ME:** "Y'ALL SAID YOU COULD BUILD ME A NEW ONE BY TOMORROW...
> WHAT WOULD THAT COST?"

> **HIM:** "ABOUT $189."

> **ME:** "SO WHAT YOU'RE TELLING ME, IF I UNDERSTAND YOU
> CORRECTLY... IS FOR $100, I CAN TAKE MY TIME, BUILD IT MYSELF,
> AND ULTIMATELY *SAVE* NINETY BUCKS?"

> **HIM:** "BUT OURS WILL HAVE ALL THE CABLES TIED UP NICELY. IT'LL
> LOOK PRETTIER ON THE INSIDE IF WE DO IT FOR YOU."

Confident my apartment isn't infested with nocturnal elves who sneakily open, review, and judge the wiring of my technology, I assured him this was a fact I could live with.

So next time, make sure to do the math. Before you start selling alternatives, bother to understand the client's goals and what cost-avoidance or return they expect. For when your solution asks them to pay more for less, it's unlikely to shake out in your favor.

Game Night Practice

Okay, so I'm one of those people who thinks through hypothetical conversations. A lot. I realize most of the scenarios will never materialize, but I nevertheless enjoy the 'just in case' exercise. As a kid, I would get really upset when — by some random turn of events — an opportunity to reference a rehearsed conversation presented itself and the other party had the nerve to go off script.

::SHAKES FIST:: "DAMN YOU FREE WILL!"

These days I've succumbed to the idea that people think for themselves, and instead apply my internal narrations to more practical things like blogging and game night. On occasion, however, I partake in this preparatory banter with friends. As was the case while cooking dinner with Calvin a while back.

Calvin told me a story about a question he got while interviewing for a job earlier that day. His soon-to-be boss asked him: if you were an animal, what kind of animal you would be? This quickly devolved into an episode of just-in-case-we're-ever-on-the-Newlywed-Game. We found ourselves likening each other to everything from wildlife to cars to plants. When it hit me! The best — albeit kind of rude — answer to the "if your spouse was a plant, which plant would they be?" question EVER:

"MY HUSBAND IS LIKE GRASS, A LITTLE BIT SLOW AND I NEED TO CUT HIM DOWN EVERY COUPLE OF WEEKS TO JUST KEEP HIM IN LINE."

Aside from the private amusement and perceived wit that

comes from having considered a comeback to every oddball remark in advance, this kind of conversational preparation finds a welcomed place in sales. Whether you struggle with objection handling or simply long to deliver an elevator pitch that doesn't require a trip to the Sears Tower[27], it might be time to practice on the little voice in your head. Just for fun.

Come up with quality, memorable responses you can deliver in a pinch. Start by answering the questions: why you, why your product, and why now. Then try to tear yourself down. This way you'll know all the holes, and all the strengths, in your arguments.

So next time, talk to yourself. A well stocked arsenal of canned pitches will catapult your sales confidence and your customer will certainly appreciate the new found conviction with which you close them.

---

[27] WILLIS TOWER* (WHATEVER, IT WILL ALWAYS BE THE SEARS TOWER TO ME.)

The Gun Show

Immediately after Ralph and I arrived at the gun show we ran into a couple of his old colleagues. I suppose finding retired ATF[28] agents at a firearm festival isn't terribly far fetched, but Ben and Kurt sure were surprised to see us. As they applauded me for "putting up with" this event, Ralph broke the news that I was actually the one who suggested the activity.

You see, I've had my concealed carry permit for a year now and figure it's about time I invest in a firearm. I mean it's not like I *need* a gun or anything; I just want to be a good shot for the sake of being a good shot − and if shooting is anything like bowling, consistency of equipment is the first step toward proficiency.

This was music to Ben's ears. Immediately upon hearing we were shopping for me, he announced:

"OH, THIS IS FOR YOU? YOU HAVE TO GET GLOCK'S NEW 380."

He then proceeded to march Ralph and me over to his preferred purveyor and instructed me to feel it out. I wasn't a fan. In fact, I'm not a fan of 380's in general. They are really small which causes the trigger to be annoyingly close to the handle. It feels like a toy gun. So I thanked Ben for the suggestion and walked him through why I didn't care for it.

But Ben didn't care.
And Ben didn't listen either...

---

[28] BUREAU OF ALCOHOL, TOBACCO, FIREARMS, AND EXPLOSIVES

BEN: "THIS IS THE BEST GUN AT THE SHOW! THEY WILL GIVE YOU A GOOD DEAL."

ME: "I'M SURE THEY WILL, AND IT IS NICE. IT'S PROBABLY THE NICEST 380 I'VE HANDLED, BUT I REALLY DON'T LIKE IT."

BEN, SPEAKING TO THE SALESMAN: "HOW MUCH CAN YOU DO IT FOR?"

RALPH: "OH NO, WE'RE NOT BUYING ONE TODAY. SHE JUST WANTS TO SEE WHICH SHE LIKES THE FEEL OF."

ME: "BEN, I APPRECIATE YOUR ENTHUSIASM, BUT THE HANDLE TO TRIGGER DISTANCE IS JUST TOO SHORT."

SALESMAN: "OH YOU CAN PUT A GRIP ON IT."

BEN: "YEA, THAT'S AN EASY FIX."

They went on for a while about this grip thing. Trying to hard sell me on the merits of an aftermarket sheath, but everyone — save for Ralph — missed the point. They insisted on solving the wrong problem. A grip would indeed fatten the handle, but it wouldn't relocate the trigger to a grown-up sized distance from it.

ME: "SERIOUSLY GUYS, YOU'RE SOLVING A PROBLEM I DON'T HAVE!"

They didn't believe me. Frustrated, I wandered off for a minute. By the time the boys caught up with me I found a few that felt nice in my hand.

ME: "THIS ONE ISN'T BAD."

BEN: "NO, YOU NEED TO GET THE GLOCK. IT'S THE BEST GUN AT THE SHOW. COME WITH ME."

He grabbed me by the hand and paraded me back over to his gun guy.

BEN: "THEY'LL GIVE IT TO YOU FOR [THIS PRICE]. THAT'S A GOOD DEAL, YOU SHOULD BUY IT."

RALPH: "THANKS, BUT SHE'S NOT GOING TO BUY ANYTHING WITHOUT SHOOTING IT."

BEN, SPEAKING TO ME: "YOU JUST NEED TO MAKE A DECISION."

This was about when I gave up. I like this guy and any further involvement in the conversation would only lead to me getting cross. But as I walked away, I couldn't help but think about what a textbook case of shit-poor salesmanship this was. I mean sure, Ben wasn't trying to sell me to pad his own pockets – he just wanted to help – but as they say, everyone is in sales and today Ben sucked at it.

How hard would it have been for him to just ask a couple questions? Is this what we all do? Do we hear a couple unrelated facts and stitch them into a web designed solely to support our preconceived agenda? Just because the prospect called us, is it safe to assume they even need software to solve their problems?

Of course not! I mean for all Ben knew I was looking for a rifle. Had he taken the time to understand my use case and preferences, I'm sure he would have paired me with a product whose features were right on target. He just forgot to ask.

So next time, listen. Take the time to perform a proper discovery and take objections seriously! When you ignore client concerns and neglect to consider their thoughts, your clients – like me – will just walk away.

Viva Mexico!

It was just before sundown when I arrived at my Mexico City hotel. Ordinarily this wouldn't be a problem, but I had to pee and it appeared that all but two of the light bulbs in my room were burnt out. What's worse, neither of the functioning fixtures resided in the bathroom. Since I severely underestimated the English fluency of Mexico City — instead of trying to invent a gesticulation for light bulb — I attempted to resolve the issue myself.

Luckily for me no one nailed down the desk lamp, and the outlet in the bathroom produced power. I deemed this solution good enough for my three day stay, took a tinkle, and headed down to dinner. Following some sarcastic remark on my part, Lucas — the new guy — admitted he thought his lights were broken too. Unlike me, Lucas speaks fluent Spanish so he called the front desk to ask what's up.

Apparently international hotel rooms have these secret magic slots by the door that you place your key into to release electricity to the lights. This particularly amused me because I actually spent a solid two minutes sizing up a mysterious 80's-car-pullout-ashtray-looking-slot-thingy on my way out of the room. Unfortunately I dismissed it after determining it wasn't going to budge and was most certainly *not* a button!

This 'damn I'm a stubborn idiot' moment made me wonder how many customer have a similar experience when trialing software. You see, my lamp relocation initiative did illuminate the important spaces in my room, and while a little unconventional, it got the job done. But if I wasn't one of those people that snarkily remarks about random things all the time, I would have left this trip with an unjustified negative opinion on my accommodations. Considering the different approaches in play: asking for help vs. figuring it out yourself, where both parties would answer yes when asked if they can see in their room — how do you decipher the truly successful from the apathetic achievers?

In my experience people will deliver the polite response the first two times you inquire. Unveiling the truth requires you look past this auto-reply and dig into the three levels of 'really?'. No one wants to burden you with their baggage, but if you poke hard enough that phrase at the tip of their tongue will slip out. It might go something like:

"HOW'S YOUR ROOM?"

"FINE."

"SO, EVERYTHING OKAY?"

"YUP."

"WHAT DO YOU THINK OF THE HOTEL?"

"WELL I THINK THEY SHOULD SERIOUSLY CONSIDER INVESTING IN LIGHT BULBS!!"

... and voilà, an adaptation of the three levels of 'why?' proves useful once again.

So next time, try to identify flailing customers before they drown. Use pointed questions to help better understand how they use the application today; not only will you improve your appreciation of their needs, you'll poise yourself to politely snatch away the shovel before they dig themselves an inescapable ditch.

Hunt for the Haggle

As a kid my friends and I would go door to door and sell stories of fake scavenger hunts to see what oddities homeowners would hand over in the name of childhood glee. We'd giggle our way home with everything from basil to Barbie shoes, before sitting around trying to come up with even more bizarre requests for our next adventure. But today – a theoretical grown up – I favor the use of logic over lies.

While attending a recent street fair I rekindled my love of pointless bargaining and decided to see what I could get for two tickets. If you're not familiar, at many street festivals, you buy strips of tickets to, in turn, buy food and drink items with. These items generally cost between four and twelve tickets. Basically it's a scam. It's nearly impossible to purchase a set of sustenance equating to a multiple of tickets in a strip. Either you go home with unused tickets or buy just one more strip and start the ticket-tetris all over again.

So while pontificating about how much that bothers me over a stick of cotton candy, my friends called me out and encouraged me to either go negotiate something or let it go.

After begrudgingly sharing my stick of spun-sugar delight, I remained unsatiated and decided to return to the cotton candy vendor with a proposition. To soften her up, I began by explaining how having two tickets is useless, and how overrated sharing is. Then I asked her how much cotton candy two tickets would buy me. She hesitated, so I

continued by offering to donate my stick back to the cause —
reasoning that it would help her recuperate costs — even
though I knew full well she couldn't use the dirty wand.
Perhaps surprisingly, my plea worked. For just two tickets, I
managed to convince her to sell me some more cotton
candy!

Friends attribute success to my bubbly blondeness or the
'girl factor,' but that's an oversimplification. I suspect my
success originates from the sheer absurdity of my requests
and concludes in my favor because of my commitment to
the cause. Software sales requires a similar commitment.
Your passion catapults your credibility; your effervescence
fuels your rapport; and your commitment drives their
change.

So next time, ask for what you need. Sure, using a new
solution may seem as bizarre to naive prospects as a bunch
of twelve year olds requesting random rubbish on a Tuesday
night scavenger hunt, but when you go into the engagement
with the same gusto you would reserve for a more seasoned
client, you will successfully sell across a much wider client
base. Be bold, get it sold.

Erin Alert

While in Bangkok, Nestor — the best company-issued travel companion you could ask for — and I tried to take a dinner cruise. It all started pretty benignly. I met Nestor in the lobby of our hotel and confirmed that I was okay with taking a motorcycle taxi. The concierge hailed us a couple of bike cabs. While we waited, Nestor reminded me of the miserable traffic we sat in on the way from the airport and said there was no way we'd make our seven o'clock launch time if we took a full-size cab.

ME: "NO NEED TO SELL ME BUDDY, YOU KNOW I RIDE... I'M GAME"

The only complication was my bag. This trip fell smack dab in the middle of 'operation grown-up lady-bag.' You see, before I left my mother convinced me to pick up some purses the next time I was in Asia, and I — not knowing I'd be on a bike when I put on this under-pocketed outfit — brought one with me to the event. No worries though, Nestor had a backpack. We simply nested my bag in his, hopped on the taxi-cycles, and headed across town.

Ziggy-zaggy, ziggy-zaggy through traffic we went. Cutting around cars and racing off lights, all while trying to hold on well enough to remain upright without crossing the 'cop a feel' line with my Thai driver. All the zig-zagging worked; we arrived at the river forty-five minutes before our scheduled departure time.

We applied our time surplus to deal hunting at the only shop nearby. While we wandered about, Nestor grew concerned that our comrades hadn't yet arrived; at half past six, he called the partner we were meeting and discovered the taxis delivered us about three kilometers south of our target destination. The race was on!

We scurried over to the hotel and asked if they would hail us some bikes. They wouldn't, or couldn't ... either way we were on our own. Off to the street we ran where, after a few minutes of failed bike hailing, we had no choice but to grab the next best thing: a tuk tuk.

A tuk tuk is basically the love child of a motorcycle and a rickshaw. It's thinner than a car, but can't lane split the way you can while riding single file on a bike. After fifteen minutes and maybe a kilometer and a half of gridlock, we had to bail. At this point, however, we had advanced to a busier street and were determined to get back on bikes. I mean we had no other choice, by now it was ten minutes to seven; we were literally about to miss the boat.

While standing on the side of the road Nestor called the partner. The plan was to have him tell the driver where to go, in Thai, so as to avoid another detour. I stuck my arm out and a bike swung over to pick me up. I sat down. While the driver spoke to our guy, Nestor got it in his mind that one of these 150cc dirt bikes could accommodate both of us.

ME: "THREE PEOPLE ON A DIRTBIKE? COME ON NESTOR, HOW'S THAT GOING TO WORK?"

My ongoing attempt to flag down a bike for Nestor — while sitting on the back of another bike — confused both oncoming bik-kies and my chauffeur. I'm not sure if it was to escape the madness or the driver realizing Nestor's plea to ride one by three meant we were in quite a hurry, but the next thing I knew, we took off! Just me and the driver. No

Nestor. But no worries, right? Nestor'll be right behind me; we got this; there are tons of bikes about; plus this time we certainly have the destination correct; it'll be fine.

So ziggy-zaggy, ziggy-zaggy through traffic I went. I kept looking back to check on my comrade, and at one point — thankfully — saw that Nestor made it onto a bike. I was feeling good about it. He's right behind me. As we zipped around traffic and approached what appeared to be our destination, I realized: Nestor still had my purse!

I get dropped off in some alley, paid the man with the change in pocket, and think:

"GOOD THING I ALWAYS KEEP MY WALLET WITH ME!"

Now there I am — standing at the end of a street, in front of nothing, with very few people around, staring at the main road telling myself:

"NESTOR IS COMING ..."

"NESTOR IS COMING ..."

"HE WAS TOTALLY RIGHT BEHIND ME ..."

"IT'S FINE, HE'S COMING ..."

...

"HE'S TOTALLY NOT COMING."

"SHIT!"

So, armed with only money and charm, I pace in a circle two or three times, realize I look like a target, and commit to direction. By now I figure the boat is long gone, but the dock is still the only logical rendezvous point, and — on principle — I want to find it. A few feet away stood a little Thai man dressed as a security guard who had been watching me meander about for the last couple minutes. I figured he was my best bet for assistance and delightfully he not only spoke English, he was able to direct me toward the dock.

I scurry down an alley, see the Sheraton, run up to the dock, and confront a large sign that read:

"IF YOU'RE LOOKING FOR THE RIVER BOAT CRUISES, THEY DEPART FROM THE DOCK ON THE OTHER SIDE OF THE HOTEL."

Crap! So I run back up the dock, jump back over the hedge-fence, and make my way around the hotel. As I fought my way through a stream of people, I regained some hope. Perhaps the boat just arrived, albeit late, and these were the disembarking passengers. But when I arrived at the other dock, only toaster-leavings of people remained. No boat, no Nestor, no partner, no nothing. After an interesting charade session with a couple of Swedes, I learned it was ten minutes past the departure time.

Now what?

I decided to regroup in the Sheraton figuring, it's a nice enough hotel, they should have Internet. My plan was to get to Google Voice, text Nestor — who I hoped was having fun on the cruise — tell him I was cool, and head back to my hotel. Unfortunately I'm no hacker. My half-assed attempt to get to the command line failed and the browser demanded credentials I did not possess. Plan C time: charm the concierge and get a temporary Internet pass.

There was a queue at the concierge desk, but I had no intention of waiting my turn. I leaned on the counter as 'New York Guy' wrapped up his request for directions to some restaurant his friend recommended. Suddenly my strategy shifted. The boys from the booth next to mine at the conference said they were going to some famous bar, I wondered if it was nearby. I also couldn't help but be attracted to the nifty business card map the concierge handed New York Guy, so I interjected:

ME: "ARE YOU BY CHANCE GOING TO THAT FAMOUS BAR?"

NEW YORK GUY: "YOU MEAN SKY BAR?"

ME: "I'M NOT SURE, THE GUYS AT MY CONFERENCE SAID SOMETHING ABOUT GOING TO A FAMOUS BAR TONIGHT ... IS THAT AROUND HERE?"

CONCIERGE: "YES IT'S CLOSE"

ME: "IS IT ON THE MAP?"

CONCIERGE: "YES, ONE SECOND AND I'LL GIVE YOU ONE"

ME: "THANKS"

...

ME: "ARE YOU GOING THERE?"

NEW YORK GUY: "NO."

ME: "WANT TO?"

NEW YORK GUY: "NO. THAT'S MY PLAN FOR TOMORROW."

Since we were heading in the same direction, and I intended to use him as an unwitting escort, I chose not to push it. After securing a map-card of my own, this rubber refrigeration-hose factory-facilitator and I walked toward my potential rendezvous point. Before I bowed out I warned him that − should my mission fail − I'd come back and join him for dinner. I suggested he order an appetizer, just in case. He smiled and I hustled onward.

The directions led me to a corner with a building on it that was in no way tall enough to house 'sky' anything. But the building across the street was very, very tall. Figuring that must be it, I headed in the only door I could find: on the loading dock. I wandered down some dark hallway, past a kitchen, and finally through a set of double doors that revealed a super fancy lobby.

Remember now, I'm dressed for a boat cruise, not a twenty-five-dollar-a-cocktail bar; I looked good from the waist up, but had capris pants and flip-flops on the rest of me. Thankfully I charmed the elevator operator into overlooking my footwear and made it upstairs.

The bar was amazing. Everything you would expect from someplace famously called Sky Bar. Beautiful views. Fresh air. Fancy people. Too bad Nestor had my camera.

Shockingly, from across the rooftop I spotted the boys. I ran over, got a drink, and told them the whole story. One of the guys had a working, data-enabled phone. He let me use it to — finally — send Nestor a smoke signal.

While I waited for Nestor's reply and sipped on my Thai mojito, I couldn't help but think about how this kind of thing happens a lot in sales too. You and your prospect embark on journey, with a perfectly reasonable plan, only to later find yourself standing all alone searching for alternate clients. Where the parallel may deviate, however, is as follows: upon finally locating Nestor I came to find he severely underestimated my resourcefulness. Fully equipped with both our phones and cameras he called everyone.

He called our hotel.

He called the Sheraton.

He called the police.

He even called the US embassy!

In under an hour, Nestor managed to get an actual 'Erin alert' issued.

But that's Nestor for you. He's not just a great travel buddy, friend, and colleague – he's a determined salesman. When I – from his perspective, playing the role of the prospect – diverged from our plan, he did everything he could to get me back on track. Now I may have since vowed to never again leave the side of the man carrying my phone, the real lesson here is that in life, as in sales, communication is key.

So next time, make a plan (and maybe a couple backup ones too) and share it with your prospect. That way should one or both of you wander off the original close path you can rest assured that both parties are committed to the goal, equipped to operate autonomously, and will – eventually – arrive safe and sound.

Sundae, Sans Cherry

Shale and I walk into the Burger King on 5th Street in Miami Beach, look at each other and conclude it was my turn to deal with the humans. I stroll up to the counter and announce my order to the lady.

> WITH WIDE, PANIC EYES SHE SAYS: "OH NO WE'RE OUT OF CO... WAIT DID YOU SAY YOU WANTED THE ICE CREAM IN A CUP?"
>
> ME: "YUP! TWO CUPS OF VANILLA, ONE WITH CARAMEL AND THE OTHER WITH CHOCOLATE TOPPING PLEASE."

Relieved she didn't have to disappoint another cone coveter, she relayed my order to the drive-thru chick. Had I known the only operable register was by the window we would've stayed in the car. But whatever, ice cream was coming — not the ice cream we ordered, but ice cream nonetheless.

We take our chocolate *and* caramel sundaes and make our way to the condiment counter only to discover it is strikingly absent of a utensil bin. So I pivot about, scurry back to the counter, and ask the woman for a couple of spoons.

> "OH, WE DON'T HAVE ANY SPOONS," SHE SAYS AS SHE SPINS BACK TO FACE HER FAUX REGISTER LIKE THE CONVERSATION WAS OVER.
>
> ME: "SERIOUSLY?!?"

Shale heard the scorn in my tone; he promptly reappeared, prepared to rescue the woman from a lecture, but she reengaged in our chat before that became necessary. Forks were about to become a suitable sundae shovel.

Seriously how do you insist someone take their ice cream in

a dish knowing full well that you have nothing for them to eat it with?!? That's like selling someone software with full knowledge that it's incompatible with their systems. Or selling an integration that's a horrible idea just because it's technically possible.

Oh wait... we do that last one, don't we?

The battle between could and should gets bloody quickly. Sales breathes down the neck of sales engineering, demanding they get out of the way of new business. Implementation gets irritated when they realize the actual effort involved in our bright ideas. Prospects expect you find their ideas as brilliant as they do. It's easy to back your sales engineer into a corner.

How do we, as sales engineers, stand our ground? How do we, as salesmen, stop bullying our brethren into blessing bad ideas? How do we, as a team, keep the client feeling like a leader?

Simple: just tell the customer the truth. A sentence as simple as, 'yes, our application comes with the tools necessary to complete your request, but it wasn't designed for that so I'd recommend you do {anything else},' goes a long way. If they choose not to take the advice, at least your bum is out of the sun, and if they do pursue the better route, everyone wins.

So next time, stock a spoon when you sell that sundae. There will always be prospects who demand to have their cake and eat it too, but don't confuse complacency with consultancy. Ultimately our job is to not just get customer, but keep them too; focus on prospects who will stick around and your stock will soar.

Be True to You

The first year I attended the Winter Music Conference in Miami I went to the International Dance Music Awards. People of all types, from every corner of the globe, gathered to honor their colleagues and comrades at this intimate affair. While in line, I made friends with a quartet of people: one Miamian, one silicon valley CEO, and two Texans. They welcomed me into the 'group' since they had — for the most part — just met as well.

As we watched the awards, I realized the thing I enjoy most about the technology field is also present in the electronic dance music community — a focus on the results, not the author's attributes. Unlike many of the professions that promise fame and stardom, in EDM[29], good looks and charming personalities are not strict prerequisites for success. Don't get me wrong — most of the people here are quite attractive, but even in the face of muscle bound men and beautifully bodied ladies, the emphasis — and the focus — remains on the work.

Throughout my career, spanning a myriad of roles and responsibilities, two facts have emerged. 1) When the company focused on the fruits of my labor, I thrived. And 2) when my boss started worrying more about my missing mascara than my work, I matriculated. I realize there will always be times for thinking and times for doing, times for opinions and times for operations, times for innovation and times for obligation. But happiness, I find, resides in these tuples' formers, which got me thinking about assumptions.

---

[29] ELECTRONIC DANCE MUSIC

We talk about our — often complicated — product or service every single day. We know the jargon, have memorized the abbreviations, and dish out business shattering ideas with all the authority of a certified industry expert. That's all well and good, but eventually, eventually we all go stupid. Eventually we commit the cardinal sin of sales. Eventually we ask our prospects to choose between embarrassment and burden. Eventually we ask them what *they* want.

That's right, I said it. I just equated asking the customer what they want to the party foul of all party fouls. Please, hang on to those pitchforks and indulge me for a moment.

I'm not saying you shouldn't discuss a customer's needs, expectations, goals, requirements, fears, concerns, aspirations, or timeline. In fact, please do! I'm saying once you've done all that, when you've positioned yourself as the trusted advisor, after you've presented a solution outline that they love — don't go muck it all up by expecting your decision maker to choose betweens technology specifics.

If you did your job well, you're having a business conversation, and your team — your sales engineers, consultants, project managers, and their counterparts on the customer's side — had the technical conversations. Since the project has the green light, go for the close; don't ask them what color to paint the door.

So next time, do what you do well. Don't try to fill every role in the deal; don't try to make DJs wear suits; don't mistake a lack of makeup for a lack of professionalism; and don't ask your customer to define *your* solution. With everyone focusing on what they do best, the deal will close as planned and on schedule.

No no ... you first!

Once upon a time, I was on a sales call with a guy — we'll call him Dale — who got deep into a sales engagement while mistakenly thinking he had the upper hand. As the buyer's journey waned on, the prospect grew impatient. Each party waited for the other to take, or at least define the next step. Every time the prospect inquired about meeting times, this 'salesman' would just turn the conversation elsewhere, often wasting time reviewing the logistics of unrelated tasks.

Then one day, as the quarter drew to a close, Dale demanded that the prospect hurry up and take action. He confessed, in a fit of urgency, that he'd been waiting for the customer to provide payment terms before he sent the discovery document required to guide the next meeting. This infuriated the prospect. Had she known he wanted the information so badly, she would have forwarded it long ago; but what really chapped her ass — as a salesman herself — was how poorly he handled the give-to-get opportunity.

Dale should have used the time between sending the discovery document and the subsequent meeting to provoke action. Each party understood that the paid discovery would not, could not, commence prior to establishing payment terms. Had he been more adroit, he would have understood that conceding to a meeting schedule would impose a deadline on his prospect, and ultimately expedite his sale.

So next time, speak up. When your customer says: "sounds good, what's next" they are handing you the reigns. Take them! When you lead the way, you close the deal.

Speed Demon

Every so often Ralph plans these super-secret trips for us to go on. The first one featured hang-gliding, something I've always wanted to try but never had the guts to do. Our visit to the Miami-Homestead Speedway to drive stock cars was another such trip. Having no advance knowledge of these trips means I have zero time to prepare, zero time to research, zero time to disclaim, and *zero* time to lower expectations. But I completely trust Ralph so I'm generally happy to trade preparation for genuine surprise.

But this is the thing: I haven't driven a stick shift car since my uncle took me out in a parking lot — ONE time, when I was sixteen. Sure, since then I've amassed several years of manual transmission experience, but it's all been on a bike. In a car you shift — if I recall correctly — with your feet, not your hands.

I spent most of the pre-ride debrief trying to convince myself that I'd be fine, that a conceptual understanding of shifting technique would be sufficient. Getting going is the hard part, that's done in the safety of the pit, and — by their own instruction — you "do not touch the clutch once you're on the track." By the time class ended I got over this particular worry-driven problem; I went off to queue for my pre-drive ride-along with the pro feeling good about the plan and looking forward to the experience.

Then, as I pulled the fire suit up over my jeans and picked a helmet off the wall, I suddenly remembered how much of a speed wuss I was. The one time I decided to 'go fast' on my motorcycle I backed off the throttle at the breakout speed of 92 MPH. I knew I would have to sack up and put the pedal to the metal today though; it's not every day you get an opportunity to go balls out on a racetrack.

As I climbed into the driver's seat and got strapped in, it hit me: this was actually happening. Nervous, I stalled it twice before finally getting the car in gear. But once I manned up and gave it some gas, I was good. As instructed, I shifted my way through the gears prior to exiting pit row and merged onto the track far more smoothly than I expected.

By the second turn I had the gas firmly on the floor — literally - and remember thinking: "this isn't as bad as I thought." The engine roared as I zipped around the track. I was having a good ol' time. Then, all of a sudden, the guy in the tower came on the radio to tell me something about my gear. The speaker in my ear was so loud, I couldn't quite make out what he said; plus being on a one-way device, I couldn't ask him to repeat the remark. After another turn or two I was ordered back to pit row.

Upon pulling in, a man strolled up to my window, lowered the safety net, looked me in the eye, and goes:

"YOU KNOW YOU WERE IN SECOND GEAR."

ME: "OH?"

HIM: "DO YOU KNOW HOW TO DRIVE STICK?"

ME: "CONCEPTUALLY."

HIM: "WHAT? ... DO YOU KNOW HOW TO DRIVE STICK?"

ME: "I HAVEN'T IN FOREVER, BUT I RIDE MOTORCYCLES ALL THE TIME. SO YEA, CONCEPTUALLY I KNOW HOW THE SYSTEM WORKS."

HIM: "YOU WERE IN SECOND GEAR."

ME: "WELL ... NOT ON PURPOSE!"

HIM: "YOU'RE GOING TO CRASH OUT THERE. I'M GOING TO SEND YOU BACK OUT, BUT YOU HAVE TO GET IT INTO FOURTH!"

ME: "ABSOLUTELY."

As he demonstrated the correct shift pattern — again — I realized how it happened. You see, the clutch was really far away. Even with the pad they placed behind my back to shorten the distance from the car's fixed seat to the pedals, I still had to lift up my ass to engage the clutch fully. Couple that fact with the very narrow left-to-right distance between the gears, and you get a situation where even the slightest tug toward the left shifted you down, not up. Turns out as I lunged and leveraged my leg forward, my weight shifted to the left and with it — unbeknownst to me — came the shifter.

Oh well.

Smoking engine notwithstanding, I managed not to die; I'm considering it a win. Yet as I climbed out of the car and debriefed with my clutch coach, I couldn't help but think about how often our prospects do dumb things during software trials. While I was well on my way to setting fire to an engine, I kept thinking: "man, I kind of expected it to go faster." If the tower guy hadn't called me back in, my actions would have caused the whole experience to blow up in my face – literally.

What about our prospects? What misguided maneuvers could cause them to develop inaccurate opinions of the product? Do we pay enough attention to how they trial our tools? Do we ask them questions that go beyond how they feel about it? Do we have any idea whether things were actually working as intended?

Probably not. A large portion of a sale's risk stems from an under-educated client zipping around an application without a guide to their goal. I was lucky; I got a pause and – because they "needed to get some air going over the engine, pronto" – an opportunity to reevaluate the experience. But a prospect who's allowed to exhaust their lap allowance before anyone assesses their efficacy is unlikely to exit the evaluation track with buying intentions.

So next time, be a coach. You've escorted your client this far, don't abandon them just because they requested the opportunity to go it alone for a little bit. When you stick by your prospects throughout the process – in person and in spirit – you guarantee they'll arrive to the close quickly and in one piece.

Scratch-n-Sniff

Someone asked me recently how I "got this way," and while I'm sure I responded with some remark as sarcastic as it was clever, the question did prompt me to reflect on how I developed into such a stubborn sell. After some thought I've come to the realization that Folgers Coffee is to blame. One clever commercial and I'd forever meet advertising ascertains with skepticism and scorn.

It all went down one afternoon when I was about nine years old. I stayed home sick from school and − fever and cough medicine notwithstanding − remember the events quite vividly. I was laying on the loveseat watching *David the Gnome* on Nickelodeon while my dad was sleeping (he worked nights) when a commercial for Folgers Instant Coffee came on TV. The ad alleged that they had just perfected new over-the-air scratch-and-sniff technology, and invited viewers to try it. Mind you, this was circa 1992, when TV screens had depth and texture all their own − very scratchable indeed.

I looked around to make sure none of the neighbors were outside and my Dad hadn't snuck downstairs. With no one eligible to witness my gullibility, turned to Sunshine − my parakeet − and said:

"GOOD FOR THEM, PILOTING THIS ON NICKELODEON BECAUSE ADULTS WOULD NEVER GET UP TO TRY IT!"

As I approached the TV, I began to feel silly and even more self conscience. So I did what I always did when I required confirmation of solitude: I waved to the back yard and smiled widely at the trees. My logic was that if someone was there, they'd think I saw them and bugger off. If no one was there, no witnesses = no shame.

After confirming I was in the clear I scurried up to the screen, gave it a good scratch, took a big sniff, and – I swear – I COULD SMELL THE COFFEE!!!

I spent the next few hours watching bad cartoons hoping the commercial would run again so I could confirm the experiment, but before it did my Dad woke up. I raced to tell him the good news. This is when my world fell apart. He didn't believe me! You – like my Dad – may choose to blame it on the fever, some deep seeded Pavlovian trigger, or an errant Wonkavision artifact, but to this day I maintain my position.

It totally happened!

True or not, the experience does raise interesting questions about the content consumers use to fill in the gaps in your stories. How do their past personal experiences impact the efficacy of your messages? Does their frame of mind, the time of day, or even a prospect's health affect the way they perceive value in your product?

I'd say so! For one, I no longer believe anything I see on TV... on principle! Moreover the Folgers experience taught me to have a solid justification for the status quo, and status' objective, ready at ALL times – regardless of how benign the circumstance. But let's think about how this plays into conventional sales wisdom...

Most methodologies encourage us to uncover a prospect's pains, to make them really appreciate how bad they have it today, and to ride that epiphany all the way to the close. But does that work on happy people and depressed folk equally well? How do we identify the line between uncovering a commitment to change, and pushing someone over the edge into the sea of futility?

I'm no shrink, so I won't pretend to have the answer (and after this story who'd believe me if I did), but I do think the issue warrants exploration.

So next time, experiment a little. Take a look in your closed pipe and see if you can uncover a personality pattern among your prospects. Knowing how your message plays to different audiences – the nerds vs. the novices, the graduates vs. the geezers, the ebullient vs. the apathetic – allows you adapt and improve; use what you learn to differentiate yourself and your solution.

Hello, Santa?

When I was six I woke up in the middle of the night on Christmas Eve to go to the loo. As I stumbled across the hall I caught a foggy glimpse of my father futzing about by the Christmas stockings. He was behaving an awful lot like Santa was supposed to. I'm pretty sure I even saw him with one of the jolly man's cookies!

The next day, while walking through the mall with my mom we passed 'Santa.' I found it curious that he'd be working the morning after his big day. So I turned to her, told her about the night before, and said:

"SO ... IS SANTA REAL?"

MOM: "DO YOU WANT THE TRUTH OR THE STORY?"

THERE'S AN OPTION?!?

This blew my young little mind. It had never occurred to me that when people asked if you believed in something — like Santa — it meant you had the option to choose not to. My mind began to race.

"WHAT ABOUT THE TOOTH FAIRY, THE EASTER BUNNY, GOD ... IS ANY OF IT EVEN REAL?"

Ultimately I chose the truth and — after promising not to tell my friends — learned that all those surprise presents were actually parentally sponsored. As I revisited this failure to further adopt the idea of Santa, I got to thinking about user adoption.

Very often software applications – like many everyday products – go purchased but unused. Is it the job of salesman to point out these risks? Should our pitch include disclaimers like: "64% of gym memberships sit on the shelf" and "over half of CRM installations fail to adopt"?

In the event we also sell a service package designed mitigate the risk, absolutely. But if not, where's the line between blissful ignorance and willingly negligent? What does letting a customer believe in idealism do to our return in the long term?

Look, in the end we are salesman. We're supposed to deliver a little fairy tale; we're expected to give the customer what they want. But unless we want to spend the next several years back-filling the void left in our pipeline from customers cancelling along the way, we should really consider renewals. Whether it's a fitness resolution, a software solution, or a gift-giving illusion – failing to acknowledge the fiction in your prospect's plan will leave your buyers remorseful.

So next time, give your clients a glimpse behind the curtain. Only when they enter an engagement with both eyes open can they really ever appreciate what it takes to achieve the fairy tale.

High Speed Service

Just moments before departing for the airport following my first un-official night in my California condo, something amazing happened! I encountered my first door-to-door salesman – or in this case, a couple of saleswomen. Both were my age, one geeky and one quite the looker.

LOOKER: "HOW'S YOUR HOME PHONE TREATING YOU?"

ME: "I DON'T HAVE A PHONE."

GEEK: "OH?!?"

ME: "I JUST MOVED IN; I DON'T HAVE ANY UTILITIES ACTUALLY."

I forgave this blatant failure to use CRM because the conversation that ensued was so amazing; I'm still a little upset I had to cut it short.

ME: "SO I NEED TO LEAVE SOON; I HAVE A FLIGHT TO CATCH. WHAT CAN I DO FOR YOU LADIES?"

LOOKER: "WELL WE'RE HERE ON BEHALF OF AT&T AND WANTED TO KNOW IF YOU'D LIKE TO SIGN UP FOR OUR FIBER [INTERNET] SERVICE."

ME: "HUH, I DIDN'T KNOW Y'ALL HAD FIBER OUT HERE. THAT'S COOL. LEAVE ME SOME INFO AND I'LL LOOK AT IT ON THE PLANE."

GEEK: "REALLY? WE'VE HAD IT OUT HERE FOR YEARS. IT IS REALLY MUCH BETTER THAN CABLE."

ME: "LOOK SWEETIE - AT&T IS THE REASON I BELIEVED CELL PHONE TECHNOLOGY SIMPLY DIDN'T IMPROVE BETWEEN 1999 AND 2009, AND SINCE MY SWITCH TO VERIZON I HAVEN'T GIVEN YOU GUYS A SECOND THOUGHT."

Neither were terribly happy with that remark so I tried to throw them a bone.

> ME: "BUT LISTEN, YOU TWO SEEM REASONABLE, SO I'M WILLING TO HEAR YOU OUT. GIVEN MY ABBREVIATED TIMELINE, HOWEVER, I REALLY ONLY HAVE A COUPLE OF QUESTIONS. 1) WHAT'S YOUR UPLOAD SPEED? AND 2) HOW MUCH?"

Of course I didn't get a straight answer. That would have been too easy. Instead, the looker asked who else I was considering, and the geek met my cable-ready response head on with a dissertation about their technology's vast superiority to Comcast. After a good ninety seconds, I stopped her.

> ME: "I PROMISE I UNDERSTAND HOW THE INTERNET WORKS ... BUT IT LOOKS LIKE YOUR COMRADE WANTS TO JUMP IN HERE. I THINK SHE MIGHT BE READY TO ANSWER MY ORIGINAL TWO QUESTIONS 'CAUSE SHE'S CLUTCHING THAT CLIPBOARD PRETTY HARD."

The looker released her death grip and read off a couple answers before trying to disguise her go-for-the-close maneuver with two additional textbook questions.

> LOOKER: "HOW MANY CHANNELS WOULD YOU LIKE? AND WHICH BUNDLE LOOKS THE BEST TO YOU?"

They both stood confused in the face of my TV-free position. (Apparently the geek doesn't know as much about the inter-workings of the Internet as she let on.) Once they realized I was serious, however, the geek offered me a workaround where I could scam the system and get basic cable without paying for it.

> ME: "WAY TO STICK TO YOUR GUNS LADY."

Judging by my tone alone they could tell it was now or never

— the hard sell was on!

GEEK: "WE'LL WORK AROUND YOUR SCHEDULE."

LOOKER: "AND WE CAN WAIVE THE $200 INSTALLATION FEE."

ME: "OKAY SERIOUSLY, I NEED TO GET GOING, BUT WHY DON'T YOU LEAVE ME YOUR CARD AND SOME COLLATERAL?"

LOOKER: "OH, WE DON'T HAVE BUSINESS CARDS. THEY ARE TOO HEAVY; THEY DON'T WANT US CARRYING A BUNCH OF PAPER. AND YOU CAN FIND ALL THE INFORMATION ONLINE."

ME: "HEAVY?"

(I CHUCKLED.)

ME: "NEVERMIND. SO GREAT, I'LL JUST SIGN UP ONLINE LATER."

LOOKER: "OH NO, THIS DEAL IS ONLY AVAILABLE FROM US."

ME: "I DON'T KNOW HOW I CAN BE MORE CLEAR. I APPRECIATE WHAT YOU'RE TRYING TO DO RIGHT NOW, BUT I'M NOT GOING TO SIGN, BUY, OR AGREE TO ANYTHING TODAY."

This didn't disrupt the geeks drive, and I swear I could see rage brewing in the pretty one's eyes. Having apparently forgot all about my airport intentions, they even offered to come by later that night if I needed more time to think. My reminder knocked little miss hard sell back onto her original platform.

GEEK: "THE TECHNOLOGY AT&T USES FOR INTERNET IS SO MUCH BETTER THAN COMCAST, YOU'D BE CRAZY TO EVEN CONSIDER LETTING THIS OPPORTUNITY SLIP THROUGH YOUR FINGERS!"

She then drew from the clipboard the world's longest order form. I took one look and said:

"Look even if I wanted to move forward today, I don't have time to fill all that out. 'Cause I need to jet, literally. So if you're telling me there is absolutely no way for me to secure this deal online, or contact you on Thursday, well then that's just unfortunate."

Looker: "Basically"

Me: "Fantastic. Thank you for stopping by, and please take the fact that you successfully introduced competition into the selection process as a win. I have all I need for today."

Had there been a table, the looker would have kicked the geek under it because you could feel the woman concocting her retort. Thankfully the sales lady managed to edge out a turn at the helm and ended the conversation. As I locked the door, I hoped that wasn't how I was during my first few stints as an sales engineer.

(I'm sure I was worse.)

So next time, manage your support team. Your technical resources have a specific role to fill; help them stay within the lines. Together, as a team, you will close strong. But remember: you're the quarterback, salesman. So control the field, control the deal, control the close.

Babes in Bikinis

You can't go to Brazil and *not* buy a bikini! My trip mates and fellow saleswomen — Kleo and Mary — concurred and with this one statement, we established Sunday afternoon's mission. I figured, if any garment is going to motivate me to shuffle my bum back to the gym, it's a Brazilian bikini. Consequently I found a suitable suit quite quickly; no need to be too picky when you're shopping for your future bottom. My girls, however, were a bit more particular.

Several shopping hours later we found ourselves slap happy and walking through the mall. Somewhere on the second level, we stumbled upon a lightly patronized shop. While Kleo and I entertained ourselves with a series of silly contests, Mary hit the fitting room — hard. Before long she emerged donning a darling little number. In it, she did a lap around the store — and scored several compliments along the way — before scurrying back to the fitting room to evaluate her remaining items.

A few minutes later, the other patrons darted out of the store and Portuguese started flying around like it was going out of style. I later learned the buzz regarded the other lady's decision to *not* purchase a bikini. Evidently she had been evaluating an ensemble similar to the one Mary pranced about in. She even secured approval from the beau — but still choose to walk away. After seeing the the runway stylings of my girl, the woman decided that — despite living on a completely separate continent — she couldn't, and wouldn't, compete.

While the girls apologized for costing the store a sale, I couldn't help but think:

"WOW, SOMETIMES FUD[30] REALLY DOES COME OUT THIN AIR."

When we think about competitive positioning, we often forget how incredibly irrational people are. Someone who really doesn't want to buy something will find any excuse to rationalize their choice to abstain. As we can see, sometimes you throw FUD just by showing up.

So it begs the question: how do we avoid being branded as running an − albeit accidental − negative sales campaign? And when we're on the receiving end, how do we know which competitor's entrances require us to take cover and which are truly benign?

They say the best offense is a good defense, but don't confuse having the plays with having to use them. Constantly blitzing will only tire out your team and cause the client to question your confidence. The key is to keep your eyes open and your intentions honorable. You can't help it if your presence upsets the competition, but can help whether or not you notice when the competition upsets yours.

So next time, remember trouble seldom parades past in a bikini. It's up to you to identify the competitors lurking in the shadows of your deal. That way you'll apply judicious defences before they distract your prospect from your well suited solution.

---

[30] FEAR, UNCERTAINTY, AND DOUBT

Just Try and Go

Remember when you were a kid and, before you headed out somewhere, your mother would insist you pee. It didn't matter how recently you had just gone, or how plausible your promise to make it till you got there was. Eventually the conversation would devolve, she'd drop a 'just try and go,' and off to the potty you'd prance.

As an adult who walks most places, I've grown to appreciate the philosophy; before I leave the house, see a doctor, or get on a plane, I do my best to make sure nothing will stand between me and whatever mission I'm about to set out on. So you can imagine my retort when an ultrasound technician insisted that I pop off to the toilet before we continue a procedure.

"BECAUSE THIS IS THE THING," I SAID... "I JUST PEED, NOT MORE THAN TEN MINUTES AGO. I SWEAR, I EVEN KNOW WHERE IT IS... FOUR DOORS DOWN ON THE LEFT."

I don't know why I was so adamant about not going. I guess I just assumed she thought I was lying. I even paused to give it a good think over after which I became even more convinced:

"I DON'T HAVE TO PEE, I PROMISE!"

She rolled her eyes, grabbed the wand, and instantly became my favorite person! Without even taking aim, she placed the ultrasound instrument square on my bladder and goes:

"SEE! THERE'S PEE IN THERE! NOW GO TRY. I PROMISE, YOU'LL PEE."

As I giggled toward the toilet, the phrase 'perception is reality' popped in my head. I hate that phrase. Don't get me wrong, I understand the sentiment and can even appreciate it under certain circumstances. But most people take it too far; they use it as a crutch. So let's be clear, more often than not, perception is absolutely NOT reality! Countless studies, court cases, grade school experiments, and ladies armed with ultrasonic sound machines demonstrate this daily. But like I said, I get the sentiment.

What I have a real problem with is when people go all blind faith and believe, without room for factual interference, that their perception is − in fact − reality. Because it's at that moment you lose the sale. Once I abandoned the idea that this woman might have had cause beyond cynicism to suggest a bathroom break − and began arguing a point merely supported by perception − progress came to a hault.

The difference here is that the lady, unlike most salesman, could prove me wrong in a way I couldn't argue with. Plus she held all the cards; nothing was going to happen until I caved. When the customer believes too deeply in their feelings, when they abandon reason, when they don't care to hear the truth, when they've already settled on a position, when you need them more than they need you, we lose − there's nothing left to fight about.

So next time, don't let it get that far. Identify when your prospect begins to believe their bullshit more than yours. Unless you sell a product that can 'prove it,' once your customers decide to disregard reality, no amount of insisting they "just try and go [with you]" will get you the deal.

Card Carrying Haggler

After ten years, countless beers, and Yahtzee cheers, the time had come. My bowling cards, now too sticky to shuffle and too mismatched to maintain, needed to be replaced. And pronto! I couldn't risk another week of bowling without the side poker game to distract me from my mistakes.

Now, if you're not familiar, bowling poker — or as it's more commonly referred: simply, "cards" — is a game where for each mark (strike or spare) you get, you earn one card. At the end of the game, the best five-card poker hand wins. And since we play with multiple decks and jokers wild, the best hand in this game is five-of-a-kind — or as I prefer to think of it, a Yahtzee.

Anyway, my first thought for a new card source was Amazon.com, but the only budget-friendly decks they had were the plain-old Bicycle red and blue kind. Way too generic for bowling because — as much as it may surprise you — bowlers will cheat at cards. Then I remembered I live in a tourist town riddled with beach shops carrying tchotchkes of all shapes and sizes. Surely in one of the myriad of markets might I find Miami themed playing cards of suitable obscurity.

Shortly into my hunt, while still establishing a cost baseline, I realized I haven't haggled for anything in a long time. I briefly wondered if I could get away with such a thing in a proper store here in the States. I mean, I do enjoy a good haggle while overseas; why not try it at home?

The first few stores offered really lame decks priced three for $12. Further up the beach I found a shop where they were three for $10, another asking eight bucks a pack, several sporting no cards of any kind, and one with a shop worker that insisted 156 postcards would make a suitable substitute.

I was just about to give up and settle for ten bucks worth of State of Florida, Chinese menu-looking cards when I remembered there was a store that's been "going out of business" for forever on the next block up. If anyone's going to let me negotiate, it's them right? I mean how else can they keep up the "everything must go" charade for five more months?

Inside I found South Beach cards – much cooler than the generic Florida ones – priced $2.99 each. As I spun the stand about searching for a third deck for the set, the lady came by to see if I needed help. Armed with an elaborate story and a captive audience, I knew this was my big opportunity!

I explained to her that I was going bowling later and needed new cards to play poker. This confused her; my plan was shaping up perfectly. I then asked:

ME: "HOW MUCH IF I BOUGHT THREE DECKS; IS THERE A DISCOUNT?"

"YES," SHE REPLIED PROUDLY, "THEY ARE THREE FOR TEN."

ME: "THAT'S WORSE!"

I laughed as I started to get louder to casually involve the other shop employee.

ME: "I MEAN I DON'T *NEED* THREE DECKS, BUT IF YOU CUT ME A DEAL I'LL GET A THIRD."

HER: "WELL, THEY'RE THREE FOR TEN."

ME: "SWEETHEART, THAT'S WORSE ..."

HER: <BLANK STARE>

ME: " ... AT STICKER PRICE THEY'RE THREE FOR NINE, WHY WOULD I PAY TEN? CAN YOU DO THREE FOR EIGHT?"

Realizing her error, she laughed but refused to take a position on the matter; she directed me to the counter lady who was now well acquainted with both my story and the math.

ME: "So the other lady said to ask you about knocking a dollar off the cards. Can you throw a local a bone? Give them to me for eight bucks?"

Averting her eyes, she laughed it off.

ME: "No I'm serious, can I get a dollar discount?"

I may have even said please as I stared her in the eye. Still giggling uncomfortably she tells me she can give them to me for $2.50 a pack.

$7.50 total?

Yea … I win.

I don't know if it was the eye contact that made her nervous or if she was just generally bad at math, but the day made me wonder what goes through the other parties' mind during negotiations. I'm generally pretty firm in my offers. Maybe because I knew I would have happily paid nine bucks I came off as hyper confident? I had nothing to lose. Maybe because I negotiate for sport and my lust to win allows me to distract them with a Cheshire grin? But my money's on the fact that most people are just out of practice, and the surprise of being thrust into a game they weren't prepared to play, tips the scale in my favor.

So next time, haggle for sport. Go to some store off the beaten path, try to finagle yourself a discount, and see how you fare. When you exercise your negotiating skills as the buyer, you'll be better prepared to preserve percentage points in your own deals.

Making it Rain

Travel enough and you'll eventually find yourself forced to stay outside your preferred points program. This was precisely the circumstance that lead to my partaking in what Westin calls a "heavenly shower." Guilt trip on a spigot is more like it!

Drawing the shower curtain reveals not one, but two, identical shower heads; not in the airline lounge, body or rain, configuration either. Both are fed from the same pipe pointed at the same spot in the tub. It looked like a watery interpretation of the western gunslingers of yesteryear - side by side they sat, aimed, and ready to fire.

Odd, to say the least.

Then I read the accompanying sign. Two long, simply worded paragraphs that boil down to:

> WE (WESTIN) GIVE YOU THIS BECAUSE *WE* ARE AWESOME, BUT IF YOU DON'T WANT TO BE AN EARTH DESTROYING, WATER WASTING, DOUCHE BAG ... TURN ONE NOZZLE OFF AND SHOWER LIKE A NORMAL PERSON.

I spent the entire shampoo process pondering whether that would even stifle water flow. Logically it appeared closing one valve would just focus the water into a single properly pressurized stream. I mean it's still fed by the same pipe, there's no reason to believe any potential aerators would really impact water consumption. By conditioner time I was consumed with sign-sponsored rage; no room for scientific curiosity remained. I didn't even bother to test my theory.

Seriously, what kind of company would choose to position their most patronizing signage in the one place promising to literally catch clients with their pants down? Did they assume a couple of customary 'rinse and repeat' cycles would make for a sufficient cooling off period? Trust me, no amount of condensation will conceal such condescension?

All joking aside, it felt like they were trying to be too many things to too many people. Just because the suggestion box contains visions of dual shower heads along side pleas for a 'greener' hotel bathroom experience, doesn't mean you have to placate both constituencies with a single water feature. And it made me wonder if we handle the variegated cast of characters that enter our deals with equally ineffective charms?

When a business user butts heads with their IT counterpart, do we pretend the same feature will sell both parties? Can we spin each component to comply? Should we even? When is it okay to take the product as a whole and stop focusing on each individual part?

At the end of the day - whether we're talking hotels, cars, or software - different people need different things. Weston could have chosen to 'be green' by not washing sheets every day and left the luxury of their lavatory intact and guilt free. Similarly, you can choose to sell to the business user and the IT guy in separate conversations. Two birds, two stones.

So next time, compartmentalize. Remember, the big picture is the icing; features are mere sprinkles. When you focus the vision, instead of forcing everyone into a single tub, you just might find the money will begin to rain on in.

Compelling Criteria

For a plethora of bad reasons that, taken together, sum to nothing more than a rationalization for 'because I want one,' I decided to get a new desk chair. As you know, office chairs fall into the category of things to which I assign highly specific requirements. For example ...

Since I like my office filled with sunshine and the sounds of South Beach — patio door open, air conditioning off — the new chair shall breathe. Make it of mesh, not leather or fabric. Sweat less, check.

Since I prefer to assume a position reminiscent of being shot into space — leaning back, base pivoted to the rear, feet up — the new chair shall be bendy. Bendy in many, many ways. Recline more, check.

Since the air in my apartment is salty — first world problem number nine — the new chair should possess more plastic parts than metal ones. Rust less, check.

Finally, since I sit at a table that's not technically a desk — thank you CB2[31] — the new chair shall have a very high, very stable up position. Sink less, check.

As I began the search, I figured I owed it to my yuppy self to check out 'good' chairs. I mean what's the point of not having kids if you don't spend the extra income on yourself once and a while, right? Plus, I've heard so much about the Herman Miller line, but have never actually sat in one. So I

---

[31] CB2 IS A HOME FURNISHINGS STORE, RELATED TO CRATE & BARREL.

stalked out a store on 16th Street that carried the line and swung on by. Not impressed. Don't get me wrong, the store is cool, but for $1,200 I expect the chair to conform to my bum like a pair of jeans after a hot bath.

It did not.

However, I did stumble on another brand, Knoll, that seemed quite comfy. The chair I was eyeballing had many, many articulation points and a back that let you turn about and lean without pinching your arm: a feature that would prove more useful if I didn't work at home, alone. But still very cool. I liked it. It would definitely be an improvement, but I wasn't completely sold on the fact that this chair was a thousand dollars better than my existing chair. So I decided to sleep on it for a couple of days.

You know how it is: no compelling event, no compelling product, no credit card compelled.

In the meantime, I go visit Ralph — who, as it turns out, knows what's what when it comes to office chairs. Well, design generally, but let's stay focused. So there I am, not thinking about desk chairs in the slightest, when I lean back in his. I reclined so effortlessly that I actually thought I might fall.

Amazing.

It turns out the sensation of imbalance and reclining delight that I was experiencing was actually the feeling of a wrench being thrown into my purchase plans. As Ralph shared the advantages of a mesh back *and* bottom, I couldn't help but wonder how often late-coming competitors disrupt an otherwise smooth sales cycle.

As the incumbent, for lack of a better term, how do we fend off competitive cherry pickers? As the underdog, how can we use the education the competition shared and time they spent to our advantage? As salesmen generally, how can we better identify the signs of someone who's just not quite sold?

Frankly none of the yuppy chairs really blew my skirt up, but they were an improvement. So absent of a better option I would have eventually − probably − bought one just to mark the quest as complete. Eventually. But instead, with a rejuvenated belief in chair potential, I expanded the search and ultimately saved myself $900.

So next time, create a sense or urgency. When you wait for the customer to sell themselves, your pipeline pays the price. Instead identify a compelling event, make a date, set a deadline, buy a chair, close the deal.

Technological Inertia

Remember the days when you vehemently opposed getting a smartphone? You hated how your husband couldn't seem to leave his alone for the duration of a meal and you vowed to forgo such hyper-connectivity in favor of actual human interaction. Remember teasing everyone about how their blackberry was simply too big for your dainty pockets? Going on to preach how the last thing you need is to give your mom another reason to criticize your purse-free lifestyle. And really, who needs that much access anyway? You're online all day, when you're out, it's because it's time to put the Internet down!

… okay, maybe it's just me …

Well, eleven days shy of midway through the two thousand and tenth year of someone's lord, I caved and got a smartphone. Begrudgingly, I'll admit: I'm hooked. But not for the perpetual email access and reliable phone call reasons you might think. I've come to realize that despite their functionality borderlining on excessive, these phones aren't themselves evil.

Okay, DUH! But many people have the same stick-in-the-mud attitude toward software. Even I – who generally advocates loudly for the excessive use of technology – found myself irrationally, passionately, and cynically advocating for the status quo. Something about turning in my flip phone made me nervous.

As technical sales people, we too soon forget how scary

change can be, but as they say: with great risk comes great reward. So when you find yourself up against a prospect whose breezed through the sales cycle, the guy who loved the demo, understands the value prop, and has the budget to buy — only to find them suddenly coming up with wildly off the wall objections at contract time, remember — they might just be afraid to change.

Remind them of the last time they took a technological leap that seemed excessive, or risky, or one that forced them to break a bad habit. Then ask them if they would ever go back to the old way. You don't see VCRs, answering machines or phone books giving DVRs, voicemail and Facebook a run for their money. So why should Rolodexes, spreadsheets and Post-it notes continue to blockade your prospects road to CRM success?

In fact, less than a week after seeing the shiny new world the flashlight app on my Droid illuminated, I've already talked my never-had-a-text-plan-in-her-life mother into getting one too.

So next time, remember — everyone is a hypocrite occasionally. You may find that once your most stubborn prospects allow themselves to try something new, they won't only improve their own processes (and your bottom line), they just might turn into your biggest advocates.

So what's your story?

Corralling hundreds of tourists around an endless flow of rum punch creates conversational opportunity like no other. And we both know — I absolutely *love* talking to strangers. So naturally I spent the bulk of my Caribbean holiday weekend floating from group to group and [not so] casually inserting myself into their discussions. Single and mingling, I made a quality amount of acquaintances. But the fun really kicked into high gear when I secured myself a conversational wing man.

A fellow salesman, this guy and I had a lot in common. For starters, we share a first name and an appreciation for the art of subtly challenging unsuspecting vacationers with deadpan remarks and a charming smile. Before dinner one night, the poolside pickings were slim so we decided to team up and take on a couple by the — currently unoccupied — performance pavilion. The approach and introductions seemed to go smoothly until we realized— much to our despair — that our targets didn't speak English.

Fear not though, a strong mix of ego and stubborn encouraged us to muscle through the conversation. But after a fantastic series of hand gestures garnered no engagement, the futility of our efforts became quite clear. Thankfully a couple of conveniently empty glasses served as a timely departure excuse. As we laughed it off and walked away, I began to wonder how often we fall victim to a similar, albeit more subtle, language barrier.

As we sling around jargon like a two year old with spaghetti, do we ever stop to confirm that our prospects have the faintest idea what we're going on about? How can we poll for understanding without sounding confrontational? Or worse, condescending?

What sometimes trips salespeople up is actually our greatest asset. It's likely the reason we migrated to the field in the first place; we have rockstar social skills. Having the ability to initiate, prolong, and manipulate conversations with people of all personality types is an art. Keeping the conversation properly grounded, however, is a science.

An introspective salesman who listens to their own spin may notice when their enthusiasm converts the conversation into an acronym laden artillery of letters firing so fast the prospect doesn't stand a chance. But for the rest of us, something as simple as checking in − by asking a well balanced blend of open and closed ended questions, and encouraging your prospects' curiosity − can make all the difference. The depth of their questions will indicate of how well they follow the conversation.

So next time, don't let your extroversion backfire. Take a moment and ask yourself whether the conversation's success stems from your forceful charm or your prospect's genuine interest in what you have to say. A customer who follows your phrases freely is far more likely to follow you all the way to the close.

In the Zone

On a Monday night, for the first time in well over a year, I played volleyball. As you might imagine, I was quite rusty. I mean I remembered how all the moves went and — even though my timing was more off than on — I remained determined to pull my weight at the Campbell Community Center's open gym.

At one point in the evening, one could argue that my competitiveness got the best of me. I took off to chase down a ball and — in a spectacular case of tunnel vision — I ran square into the bleachers! It was like something out of an episode of Looney Toons.

Bump, bam, bounce, backfire.

I honestly can't tell you if I made the play or not. Nor can I can say for sure whether the shock etched into the faces of the other eleven players resulted from the accident itself, or how quickly I popped back to my feet and started laughing. But as I showered off the night and browsed the brewing bruises, I couldn't help but wonder how often this type of hyperfocus negatively affects a sales cycle.

We've all encountered someone with 'happy ears' − that guy who always seems to hear what he wants, not what actually is. You know, the salesman who − no matter how obvious the signs might be to the rest of us − still never seems to receive the signals his prospect sends him. He's the guy who commits deals to his forecast that don't even deserve a quote; the gent that tries to drown objections in discounts; the comrade who wants to win so badly that he can't even imagine a loss.

Nobody wants to be *that* guy.

So how do we snap out of our daze and tune back into reality? Because the rest of the world − be it in the form of negative buying signals, or as in my case, a literal wood wall − doesn't cease to exist just because you forget to look for it. We talk a lot about listening 'between the lines' and digging for what the customer really means and needs, but if we are to succeed in sales we need to make sure we see what's right in front of us as well.

So next time, open your eyes (and your ears). Your prospect's word choice contains much more information than we often give it credit for. When you take the time to consider the whole conversation you'll surely avoid an unexpected dead end.

That's Alright

The second time I headed to Australia for work I had the opportunity to spend time in several cities. During the course of a week I managed to accumulate quite a few quips about the experience. When I first considered the following stories, I found them nitpicky and not *Sapient* worthy, but after snagging some time for rest and reflection, I realized that the series of escapades shared a common theme: passive aggressive and slightly apathetic customer service. Since nothing kills a deal faster than poor people skills, let's review...

I first experienced the 'that's not my job' attitude checking in for my flight from Sydney to Brisbane. When the OneWorld Emerald rope maze — a.k.a. the airline status line — suspiciously dead ended with no agent in sight, I wandered over to the first staffed counter I saw. As the agent's eyes rolled up to meet mine, she informed me that I should have went right, not left. Fine, no worries.

So I went to queue up behind a couple people who had been there since I entered the terminal. After patiently waiting their turn, however, the agent said:

> "YOU KNOW YOU CAN DO THAT (PRINT A LUGGAGE TAG) FROM THE KIOSK. WHY DON'T YOU GO TO THE MACHINE?"

Apparently she could have helped them, but couldn't be bothered.

Jump forward a day. Nestor and I — being the fiscally responsible employees we are — try to get a rental car to attend an early morning meeting just outside of town. The conversation with hotel staff went something like this:

NESTOR: "WE NEED TO PICK UP A CAR FROM THE AIRPORT QUITE EARLY TOMORROW MORNING, AT LEAST BY 8, BUT I HEARD FROM OUR MAN ON THE GROUND THAT THE AIRPORT HERE MIGHT CLOSE OVERNIGHT."

DESK AGENT: <NO REACTION>

ME: "DOES THE AIRPORT CLOSE?"

DESK AGENT: "YES."

ME: "WHEN DOES IT OPEN?"

DESK AGENT: "BEFORE THE FIRST FLIGHT OUT."

NESTOR: "DO YOU KNOW WHEN THAT IS?"

DESK AGENT: "NO."

ME: "OK! SO..."

NESTOR (CUTTING IN): "LET'S THINK THIS THROUGH... WHEN'S THE EARLIEST ANYONE HAS CHECKED OUT TO CATCH A FLIGHT?"

DESK AGENT: "05:00, FOR A 06:05 FLIGHT"

With such incredibly precise, borderline obstinate, responses you'd think he was testifying in a capital trial, not providing customer 'service.'

But my favorite quote came in the form of exiting wishes offered by a clerk during a brief shopping excursion. For the record, I've always found stores where staff meet direct eye contact with complete silence quite hostile. The glares remind me of how parents watch kids who they expect are fixin' to misbehave, and consequently I feel guilty for even thinking about touching things. Needless to say, I seldom purchase anything from such shops, but I always say "thanks" as I leave.

Usually I mean, "thanks for letting me look," but times like this it's more "thanks for not yelling at me." Either way I understand that I'm a guest in their shop and try to acknowledge that fact.

So anyway, I'm walking out of this one beach shop on the strip in Noosa Heads and say:

"THANKS!"

LADY: "THAT'S ALRIGHT"

"That's alright"?? You got to understand, she didn't say it in a 'no worries' kind of way. It was more in a sardonic 'it's okay that you didn't bother to buy anything, don't feel guilty, I had nothing better to do than to watch you look around,' sort of way. Charming.

Even hours later Nestor was still fuming over this woman's tone. So to lighten the mood, we discussed the myriad of customer service stunts witnessed to date over a couple of beers. Jokingly reminiscing about the painstakingly slow pint pours and snarky cabin crew, Nestor and I drew the attention of an Aussie who graciously explained the cultural roots behind this friendly behavior.

Apparently — and to be clear, these are his words, not mine — Australians in the service industry are not a fan of Americans. The mere fact that we refer to the 'service industry' as such pisses them off. We came to learn that the minimum wage down under is quite high, making tipping not customary, and the fact that Americans insist on tipping is an insult.

"WE'RE DOING OUR JOB, WE GET PAID WELL FOR IT, WE DON'T NEED YOUR CHARITY."

He went on to explain that they see themselves as equals, not "servants," and don't wish to be pandered to. He believed what we had been experiencing was posturing designed to level the power playing field and remind us of this equality.

To be honest, I found the exchange more rationalize-y than reasonable. It's not like either Nestor nor I walk around bars with an attitude like we expect to be waited upon and pandered to like pompous rich folk. And — for the record — shopkeepers, waitresses, and customer service agents *are* our equals; it's only after people act dickish that my respect for — and behavior towards — them changes.

Either way, the lesson lays in the effects, not the cause.

This is one of those times where the "service is the new marketing" mantra really applies. It doesn't matter if you think your customer needs to be taught a lesson. Because like it or not, if you're 'customer facing' you're the face of the company. As Nestor put it:

"I CAN BUY A T-SHIRT ANYWHERE, I DON'T NEED TO GO WHERE I'M UNWELCOME."

So next time, swap the FML[32] face for a smile. Even if the grin doesn't erode your passive aggressive tendencies, it certainly makes them more palatable. Whether you're having a good day or bad, harness the power of positivity; send them off with a smile.

---

[32] FML - Fuck My Life

Tazmania!

After ten days and four cities, my latest Australian business trip was drawing to a close. I had one more stop left and I was really excited about it. I had never been to Tasmania before, so when a meeting ran short, I squeezed onto an earlier flight to Hobart. Fueled by Tasmanian excitement I went around telling my fellow travelers about this next and final stop on my journey. Unfortunately my single serving sounding-boards responded with under-tailored retorts.

Everyone kept telling me stories about how beautiful Hobart was. When I asked about things to go do or places to visit during my one-day stay, they deflected and returned conversational focus to the serene landscape. People kept assuming a well-hyped backdrop would perfectly frame my happiness. After a half a dozen stories about the majesty, the charm, the modesty, and the quaintness of Hobart, these Aussie travel mates had nearly extinguished my excitement. Yet I told myself that − while their assertions may be well founded, they certainly weren't comprehensive; I actively maintained my optimism.

I deplaned by stairs and Tarmac, followed a walkway to the one and only arrivals gate, and abandoned my airline-issued orange at the 'recommendation' of an angry beagle. Breezing through the door, I basically tripped over the baggage claim which curiously had a statue of a sea lion riding around in circles on it. So far, pretty quaint indeed.

The ride to the hotel came complete with beautiful backdrops and charming conversation − nice and all − but I couldn't help but notice the severe lack of stuff along the way. There were no buildings, no businesses, and no bustle of any kind. By ten minutes past five o'clock in the evening I was checked in, changed, and ready to explore.

ME: "WHICH WAY SHOULD I HEAD TO FIND THE BEST SHOPPING?"

HOTEL LADY: "OH, IT'S AFTER FIVE, NOTHING IS OPEN. WELL... TARGET MIGHT BE OPEN UNTIL 5:30 IF YOU HURRY."

This made me smirk, which apparently insulted her, as it garnered a follow on remark.

HOTEL LADY: "THIS ISN'T ONE OF *THOSE* KIND OF CITIES."

It turns out she is right — although not about Hobart being a city; it's a town at best. Disappointed, I wandered around in the cold for a couple of hours. My stroll confirmed that after sunset — in a city absent of street lights — there's not much to see. As I moseyed back to my room to order grub from the only open kitchen in the area — room service — I couldn't help but wonder if my disappointment was exaggerated because everyone kept pointing out the things about the place that I wouldn't like?

Had someone taken a moment to discover that I'm not the kind of girl who finds parochial charming and used it to instead highlight the great people and presumably good food, would I have neglected to notice the limited evening activities? I'd like to think so.

So next time, discover *then* sell. People like, and purchase, things for very different reasons. When you fail to appreciate that fact you risk tainting their experience and forever jeopardize their ability to be objective. Do some discovery; tailor your spiel; win the deal.

Strategically Stubborn

To this day I am continually surprised by how often I find myself in conversations concluding with everyone in 'violent agreement'. You know, when you agree with the other guy, but for some reason both of you are yelling. But what's really frustrating is that days — if not hours — after these discussions terminate, the behavior demonstrated by the other party bears no resemblance to the practices they so passionately preached. But today, I think I figured out why!

Salesmen, especially the successful ones, share a unique mixture of stubborn, habitual, and self-centered behaviour. This trifecta — arguably the successful symptomatic subset of full-blown narcissism — allows us to roll with the punches, persevere amidst defeat, and blow out our numbers quarter after quarter. These qualities, genuine personality traits as they are, don't limit their manifestations to the workplace. They're with us each and every day. Which incidentally, is what prompted me to arrive at this little epiphany.

Despite still boasting a broken ankle, I decided to escape for a weekend. I arrived at MIA with a plan: I would hobble my way to the skycap, ditch my luggage, and gimp over to TSA. The skycap grilled me about knowing how to get to my gate and pushed HARD for me to use a wheelchair. But I was really looking forward to pre-checking on crutches; the opportunity to graduate from the 'dump and go' to the 'drop and hop' filled me with some hard-core traveler pride.

So, with my sights set firmly on the impending TSA experience, I resisted plea his with a stupid story being a highly experienced three-legged racer. You know, "I Crutch for the Cure" and all that jazz.

After an eye roll, I was released to TSA who also offered to summon a steward. This only strengthened my will. I hopped away proud and Murphy[33] took notice. My plan was to crutch my way to the Admirals club, call it a workout, and consider a chaperone the rest of the way. As a fully functional biped I never gave the distance from curb to club much thought. Not only did I actually work up a little bit of a sweat during this .42 mile jaunt, I had a lady stop me along the way.

Apparently my laptop bag had been slowly eating my dress. I have no idea how long I swung along with my ass hanging out before my wardrobe's malfunction was brought to my attention, but it honestly didn't phase me. Peep show notwithstanding, I was keeping up with people, and *that's* what really mattered.

---

[33] OF MURPHY'S LAW

Stubborn? Check.

During my gallop I noticed the 20:00 flight was delayed. Autopilot kicked in; I always try to get on the earlier flight — mostly because I can, using my status to jump the standby line makes me feel special. So upon arrival at the Admirals Club, I immediately asked to go standby. This was stupid. My upgrade had already confirmed on the 22:05 flight, I checked a bag, and I really needed to stop sweating and start drinking before escorting my broken bones skyward.

Habitual? Check.

Luckily the lady talked me out of it, but not before explaining to me — in detail — why I couldn't have the exit row seat. I suppose she didn't realize I fly enough to have the safety videos memorized. For me it's like a song that occasionally gets stuck in my head. Especially that one line that salesmen everywhere can embrace as validation of our otherwise universally frowned upon behavior. When it comes to oxygen "always put your own mask on first...". Look out for number one; you don't hit quota by suffocating your pipeline or stifling your efforts by helping others before you help yourself.

Self-centered? Check.

So next time, don't fight your instincts, but do question your habits. Those violent conversational agreements might just be your subconscious gently advising you against a flight path that won't actually advance the pursuit of your larger goals. When you apply your will to a more modern method, you'll certainly soar past the competition.

Utilities

The day after my run in with the door-to-door saleswomen who tried to sell me Internet, I arrived in Germany almost ready to work. First, however, I needed sleep. Lots of sleep. After a thirteen hour jet-lag coma-nap, I woke up bright-eyed and ready to train some salesmen.

Then I looked at my phone. On it? A text message from my painter that read:

> "JUST A HEADS UP SOMEONE CAME AND TURNED OFF YOUR POWER AND WATER TODAY, DON'T WORRY THOUGH I TOOK YOUR FOOD HOME WITH ME. JUST LET ME KNOW WHEN IT'S BACK ON SO WE CAN FINISH!"

Even half awake I correctly surmised that it was still August 1st in California and − if memory served − Fernando, the landlord's broker, said not to deal with the utilities until the first. Plus where I'm from utility companies don't shut off the electricity the day someone moves out. It's the unspoken courtesy rule of rentals, utilities stay on and the next person to move in pays for the ninety-two cents of power that was used to heat and illuminate the unoccupied place.

This situation is better for everyone. Real estate agents have lights to show the property with. The pipes don't freeze and burst in the cold. The fridge isn't stinky. And the electric and gas companies don't have the expense of sending people all over to constantly turn switches off and on again.

What's the deal California?!? Why so mean?

Needless to say, it took little further nudging to rocket me from zero to feisty. Much to my delight said nudge awaited me in text message number two. This one from Fernando, who wrote to report he successfully scheduled the gardener, electrician, and carpet cleaner to come the next day. Unable to imagine a set of three contractors more in need of electricity and water, I fired back quick notes, and turned my attention to PG&E[34].

I called their help hotline and spent two or three minutes in queue before briefing the agent on the situation. The agent found my little rant amusing — which temporarily defused some of my rage — and set up my an account. "So far so good," I thought ... then he started laughing. Laughing like I do when I'm heckling someone in my head.

ME: "WHAT'S FUNNY? NOTHING'S FUNNY? WHY ARE WE LAUGHING?"

Sure enough with his regained composure he regretfully informed me the two utilities in question are provided by the City of Santa Clara directly, and they only work from 8am-5pm PDT. Such remarkably convenient operating hours provided me a much needed twelve hour festering period.

At 08:02 PDT on August 2nd, the conversation, that began with me civilly explaining the situation, ended something like this...

LADY (AS SHE COLLECTS MY INFO): "THE EARLIEST WE CAN GET SOMEONE OUT IS TOMORROW."

I launched into an editorial about how utterly ridiculous I found the situation, followed by a brief history lesson about

---

[34] PACIFIC GAS AND ELECTRIC

how utilities work in the more accommodating parts of the country, i.e. Chicago, and resubmitted my suggestion that they be less eager to cut the power and maybe adopt some sort of a grace period.

LADY: "WE'RE NOT ALLOWED TO GIVE GRACE PERIODS IN CALIFORNIA, AND TOMORROW'S THE NEXT AVAILABLE APPOINTMENT WE HAVE."

I opted not to dive into the legal merits of "not allowed."

ME: "BY YOUR OWN ACCORD, THE SERVICE WAS PAID THROUGH THE FIRST OF THE MONTH, SO HOW DOES CUTTING POWER BEFORE NOON, ON THE FIRST, NOT SEEM RIDICULOUS?"

LADY: "I'M NOT SURE WHY THINGS WERE SCHEDULED THAT WAY, BUT I CAN OFFER YOU AN APPOINTMENT TOMORROW. WOULD YOU LIKE IT?"

ME: "LOOK, I APPRECIATE WHAT YOU'RE SAYING TO ME, BUT LIKE I SAID I'M IN GERMANY AT THE MOMENT AND I HAVE A FLEET OF CONTRACTORS LINED UP TO SERVICE THE UNIT IN MY ABSENCE WITH NO WAY TO RESCHEDULE.

I REALLY NEED YOU TO APPRECIATE DIRENESS HERE. THE FACT THAT I'M SPENDING TWO DOLLARS A MINUTE TO HAVE THIS CONVERSATION SHOULD BE AN INDICATION OF THE URGENCY OF THIS SITUATION.

SO SERIOUSLY, PLEASE, CALL WHOEVER YOU HAVE TO, BUT I NEED YOU TO TURN MY POWER BACK ON *TODAY*!"

She placed me on probably a ten dollar hold, but came back with good news — both my power and water would be restored by 2pm that day.

Now I don't know what's worse, the fact that I *can* manhandle phone salespeople like that, or the fact that I so frequently *need* to. Either way the pride of the conversational win does not wash away the bad taste I have for the company. Those of us not literally holding the monopoly on power will surely find a prospect going across the street before fighting for their light like me. And it got me thinking about the things we withhold — on principle — from our prospects.

Access, discounts, documentation, terms, conditions — we manage access to many of the things our clients need. So how can we tell when we're withholding to further a goal and when we're just flexing our muscles for flexing's sake? On the flip side, how do we fight for our customers without exhausting our resources?

When it comes to giving, make sure you get something too. But when you have everything you need (save, perhaps, for the final contract) putting up roadblocks only slows the deal's progression. Plus constantly bluffing makes you look weak — if you going to draw a line, make sure you're ready to stay on your side of it.

So next time, know when to hold 'em, and when to fold 'em. I'm not suggesting you suspend a quid pro quo policy in your sales negotiations, but the line between firm and infuriating is both thin and tenuous. Walk it carefully.

Parenting Prospects

My wildly underdeveloped maternal instinct manifests itself in odd ways. When I first purchased my Roomba – a floor cleaning robot that I affectionately named Glenda – I morphed into a total 'mom.' As she scurried under my couch I worried she would get tangled in the cables, or worse, trapped in the milk crate that used to hold up my lamps. So I literally followed Glenda around, tidying ahead of her arrival. Thankfully, I quickly learned what she could and could not handle.

Before her next go, I made the necessary adjustments to the room's configuration, and after her highly successful run – not only am I totally hooked, I'm a little inspired. While Glenda rid the living room of errant bird seed and feathers, I took on the cardboard cluttering up my kitchen, making this the most successful joint cleaning effort my house has ever seen.

As I basked in the glow of my shiny and clean abode, I started to think about micromanagement. Successful sales requires a lot of trust between the rep and the client. Prospects must perform many activities independently; perhaps most importantly, they will likely need to sell your solution – on your behalf – internally. How can you best prepare them for this quest, without making them feel like they're being 'mothered'?

Whether you sell ads for a daytime soap opera, a SOAP[35]

---

[35] SIMPLE OBJECT ACCESS PROTOCOL: A TYPE OF WEB-SERVICE. OR – IN LAYMAN'S TERMS – A WAY FOR TWO SOFTWARE APPLICATIONS TO TALK TO ONE ANOTHER.

service, or actual soap, I'm sure you've got a strategy to navigate this stage of the sales cycle. Maybe you provide collateral or a presentation for them to use? Perhaps you insist customers practice their pitch on you first? Whichever the case, at some point along the way, you're going to have to cut the cord.

So how do you know when your customer's ready? How can you be sure they are putting your best foot forward? When is it safe to trust your quota to another?

I'd like to believe that a true advocate needs little coaching. Little — as in some, not none. If you've successfully conveyed value to your client, and secured them as a champion, you don't need to worry about their tone or spirit. Simply provide them with all the facts they require to promote your product up the chain, and they'll do just fine.

On the other hand — if you find yourself feeling like you can't let go, if you notice yourself sheepishly following your client's every move, realize it's likely because you don't have a real champion. And that's bigger problem you need to address before you can move forward with the deal.

So next time, trust your advocates. Like Glenda who has proven she can fly solo, when your prospects demonstrate a desire to spread their wings, let them soar. When you trust your prospects, you can trust your forecast.

What's in a name?

For the benefit of the reader I've done my best to briefly provide context for all the names I've dropped throughout the book, but in real life I'm seldom so courteous. When I tell a story to my friends I say things like "the boys" to refer to whatever group of guys were most prevalent during the relevant timeline, and I prefer to just use someone's name than their byline. I mean let's be serious, if you had to pick between saying "Nestor" or "my travel buddy that ran APAC sales for Saccharin a while back" every time you referenced the character, you'd choose "Nestor" too.

Seperately, as you know, I have been enamored with Glenda's performance for a while now. I talk about her all the time. Then one day she scared the crap out of me. I was in the shower, shampooing up a storm, when she wandered into the bathroom. The problem being, however, that I completely forgot I set her out to clean. So when she cruised into the room, I about jumped out of my skin.

I thought I was being attacked! In the blink of an eye I concocted a whole theory about how I neglected to lock the front door [again] and some shady delivery man decided to come in and take advantage of me. It wasn't until I scanned to room for something I could use as a weapon that Glenda finally reached a wall, made a sound, and allowed me to see her.

Relieved, I laughed and went on with my day. But hours later, while grabbing a cocktail with Ralph, I still felt pretty stupid for getting so startled. I decided to tell him the story.

ME: "Man, Glenda startled the crap out of me while I was in the shower this afternoon."

RALPH: "How?"

ME: "I forgot she was out and about."

RALPH: "She came into the bathroom?"

ME: "Yea, and I caught her out of the corner of my eye. Damn near jumped out of my skin when she scurried past."

RALPH: "Does she have a key?!?"

ME: "What?"

RALPH: "Like to the apartment."

ME: "What are you talking about? Why would Glenda need a key? How would that even work?"

RALPH: "Isn't Glenda your cleaning lady?"

I laughed for about five solid minutes before Ralph — now firmly convinced I was laughing at him — finally stopped me to ask what was so funny.

ME: "You know I have a Roomba right? The robot sitting in the corner by the couch"

RALPH: "That thing in the corner is a vacuum?"

ME: "Yea! That's Glenda!"

RALPH: "You named your vacuum?!?."

ME: "Of course I did, who *doesn't*??"

RALPH: "This whole time I thought you had a maid."

We still laugh about this all the time and when we do it makes me think: what a classic example of what I call the 'clash of nouns' problem. When I switched from selling CRM to selling contact center software this happened all the time. In CRM a 'contact' is a person with whom you are or should be doing business. In the context of the contact center, a 'contact' is an interaction, an event where a person contacts the call center.

When we use the same words as our clients, how can we be sure we mean the same thing? Perhaps more importantly, when we try to map our prospects' bucket of nouns to the titles in our technology, how can we be sure we're assigning appropriate meaning?

Unfortunately, this is one of those problems whose solution requires subtlety. If you've developed a good relationship with your prospect, you will hopefully be able to detect when you two are no longer talking about the same thing. But if we can learn anything from my discussion with Ralph it's that − regardless of how close of a relationship you may have − once both parties are committed to their position, a lot of time may pass before a 'clash of nouns' corrects itself.

So next time, listen up. When you ask yourself whether your client's responses jive with the conversation that *you* are having, you'll catch discrepancies and correct course. Pay attention, it'll pay off.

Pick Your Battles

So I'm in a cab on my way home from the airport sporting the voice of an elderly chain smoker when I realize this marks the ninth time in as many months that I've lost my voice on the road. My current theory is that the frequent strain may be a function of limited mandibular range. Since I can't open my mouth wide enough to 'create a cathedral of sound,' I instead force my voice out via sheer shrill will.

Ajax, in between barking directions on my behalf to the driver, vouched for the plausibility of my hypothesis and suggested I try lemons. Then, in an attempt to rectify my obvious confusion, he explained:

AJAX: "MY COMMANDING OFFICERS IN ROTC[36] USED TO FREQUENTLY PRESCRIBE THE LEMON AND LIME REMEDY TO PEOPLE WITH SIMILAR AILS."

ME: "WHAT EXACTLY IS THAT SUPPOSED TO DO?"

AJAX: "THAT'S NOT SOMETHING YOU ASK."

ME: "HOLD ON A MINUTE... SOMEONE TOLD YOU TO 'GO SUCK A LEMON,' AND YOU'RE RESPONSE WAS TO HIGH TAIL IT TO THE STORE AND GET YOURSELF A LEMON?!?"

AJAX: "PRETTY MUCH."

You see where I'm from 'go suck a lemon' was the grade school equivalent of 'go fuck yourself,' so I found this utterly hilarious! But never one to belittle [out loud] without a solid scientific platform, I directed my phone's final battery power toward debunking this advice. The best part of the next ten

---

[36] RESERVE OFFICERS' TRAINING CORPS

minutes was how Ajax rationalized the remedy. As I Googled on, I realized the results' importance paled in comparison to the process in play here.

When someone believes something that's clearly wrong, is it better to correct their falsity or play into it? We all know which I pick, at least in my escapades. Yet in sales strong opinions, right or wrong, can actually play into your agenda. Like a student who realizes early on that they can get a 'B' with a tiny fraction of the effort required to garner an 'A' – a sapient salesman directs passion's inertia toward their products.

So if your customer thinks computers are the manifestation of many magical elves, put on this planet to keep the impending uprising of the apes in check. Use it! Explain to them that you sell the most magical computers in all the land, and – to your knowledge – absolutely no monkeys have advanced their mission using your tool. Hell, go onto say (what is likely true), that you have not seen or heard your competition EVER make such a claim.

The thing is, people like being right. There's no harm in allowing them to maintain their mantra. If you're quite confident your tool will address their other, more relevant needs, let it be. As long as you preserve a base level of ethics and deliver your indulgences with adequate deadpan, things will work out.

So next time, pick your battles. Many hurdles lay between your prospect and prosperity. Focus your efforts on solving business problems. When you allow yourself to ignore irrelevant, erroneous impressions, you increase client confidence while simultaneously fostering mutual success.

The Ultimate Meat Bowl Spoon

On my way to visit Ajax and Perl out in Idaho, I took a flight from Fort Lauderdale to Phoenix and sat next to a really fun dude. Andy and I yammered away for about an hour before the US Airways staff served us dinner. By now it's no secret I'm a flatware kleptomaniac[37], but Andy had no idea what crime he was about to become a party to[38].

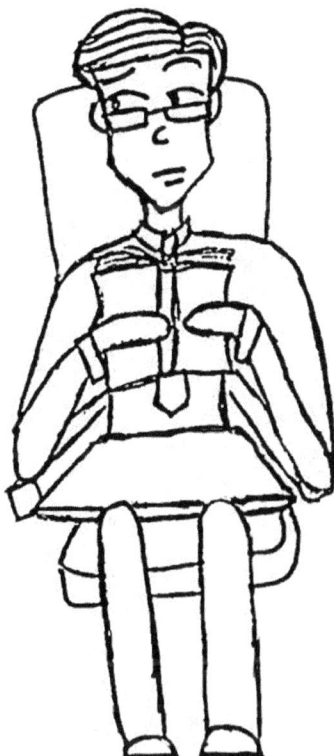

As a part of American Airlines' merger with US Airways and their journey to rebrand as the "New American", the utensils on flights needed to be replaced with ones sporting the new American Airlines logo. Not surprisingly, rollouts of new stuff is appearing on USAir flights faster than American Airlines planes. Not a big deal or anything, but having grown up as a traveler dining on AA's legacy silverware, you can imagine my surprise — and delight — when the fight attendant handed me the ULTIMATE MEAT BOWL SPOON on this, not quite American, flight.

---

[37] FOR THE RECORD, I'M TOTALLY A LAW ABIDING CITIZEN.
[38] ::LOOKS AROUND:: WHAT?!? NO CRIME HERE!

Andy was about as confused about my excitement as you probably are right now. So let me take a step back and define meat bowl for you.

Many many months ago — long before any romantic rigamarole — Ralph and I stopped by my place for a mid-bender break; I desperately needed to eat something before we continued on to meet our friends at the next bar. So I made what I usually did when in a hurry: tuna. Not the way my mom made it with eggs and mayo. No. Way better than that. With peas, corn, hot giardiniera, and hummus — my mushy masterpiece remains the perfect way to protein in a pinch.

Up until now, however, most of the meals Ralph and I shared were borderline yuppy. Even the homemade ones were plated and presented like a pro. Despite his constant claims that he "just eats ground beef and veggies" most nights, I had seen little evidence to support this claim. I was beginning to disbelieve it. But in this moment — me drunk-shoveling tuna into my face like a starving astronaut recently returned to Earth and struggling with gravity — that Ralph conceted access to his meat bowl.

> **RALPH**: "NOW THAT I KNOW YOU EAT SHIT LIKE THAT — WHICH IS DELICIOUS BY THE WAY — I'M OKAY WITH MAKING YOU SOME MEAT BOWL."

I explained to Andy —

> "NOW WE EAT MEAT BOWL FOR DINNER ALL THE TIME. BUT THIS IS THE THING — HE USES A SPOON AND UP UNTIL NOW I'VE ALWAYS USED A FORK. I'M MISSING ALL THE MEAT JUICE! DESPITE THE EXTENSIVE SELECTION OF SPOONS IN OUR HOUSE, NONE OF THEM STRIKE THAT BALANCE BETWEEN SHALLOW AND SHOVEL, BETWEEN PLATFORM AND PIT, BETWEEN SKEWER AND SKEWED — NONE ARE THE ULTIMATE MEAT BOWL SPOON."

As he laughed at my monologue I couldn't help but think about the lengths people go to to procure the perfect product. Beg, borrow, and steal — as they say, right? So in software, where nothing is perfect out of the box, how do we get people inspired to fly with us? Can passion persist without perfection?

I certainly hope so; after all, most of our careers depend on it. Luckily, in all practicality, the customer doesn't have to be the one who's excited — at least not out of the gate. As long as someone in the deal is delighted by the product, you've got room to maneuver. Excitement is as contagious as a yawn; which are you passing on?

So next time, go after what you want. If you're not passionate about the product you sell, maybe it's time to move on. When you believe in what you sell, you'll sell more than software; you'll sell solutions; you'll sell substance; you'll sell something special.

Honesty and Plausibility

Late in the season one summer I was summoned to Dallas to participate in a sales training seminar. The entire central region sales team gathered in a hotel conference room and performed various activities. One assignment asked us to come up with a "plausible emergency" we could use to scare prospects into demanding a feature of our offering that we, uniquely, provide.

Everyone did alright, mostly choosing to hate on hurricanes and play the disaster recovery card. Which is fine, but I got the feeling Jani – our workshop coordinator – had become a little too spoiled by the willingness of folks to accept 'plausibility' in a workshop setting.

So day three rolls around and Jani announces that they've reserved the adjoining conference rooms for teams to utilize during breakout sessions; my team was assigned one such room. We neither wanted nor needed a room though. Team Tactful – as we called ourselves – used sunlight to illuminate our bright ideas; we were perfectly content working from the chaise lounges, poolside.

Jani objected. She had been very uppity about our anti-paper, quietly-collaborative working styles all week and – upon seeing us begin to file outside – declared:

"YOU CAN'T WORK OUTSIDE TODAY."

ME: "I BEG YOUR PARDON?!?"

JANI: "THE HOTEL COMPLAINED AND SAID WE COULDN'T HAVE PEOPLE WORKING IN THE COMMON AREAS. SO I NEED YOU TO WORK INSIDE"

ME (FLABBERGASTED): "WHAT?!? PLEASE CLARIFY."

JANI: "THE HOTEL GUESTS FOUND THE EASELS DISTRACTING. WE RECEIVED COMPLAINTS. I CAN'T HAVE YOU SET UP OUTSIDE."

Mind you, my team used Google Docs and laptops, not tripods and paper.

ME: "SO LET ME GET THIS STRAIGHT, YOU'RE TELLING ME THAT HILTON GARDEN MANAGEMENT IS FORBIDDING HOTEL GUESTS FROM CONGREGATING IN HOTEL COMMON AREAS?"

JANI: "YES"

ME: "REALLY? THIS IS COMING FROM HILTON... *THAT'S* THE STORY YOU'RE GOING WITH?"

JANI: "I JUST NEED YOU TO TAKE YOUR TEAM TO THE CONFERENCE ROOM AND STAY INSIDE."

I didn't bother to fact check Jani's story because I literally couldn't stand to quibble with this woman about trivialities anymore. And you and I both know her story was bullshit anyway. That said – and all bickering aside – the week raised several questions regarding the risks of prefabricated excuses.

How plausible must an emergency be for a prospect to interpret it as danger and not as salesmanship? Are you prepared to stick to your guns at all costs when a conversation goes off script and a customer challenges the likelihood of your emergency? At what point should we relinquish the position and concede the tactic?

Instinctively I stick to my stories like toilet paper on a wet bottom, but if I get called out — especially when I know I'm full of shit — I've learned, at least in sales — it's best to stop, concede, and reset.

So next time, define your exit strategy. When a sales tactic goes awry, are you going to throw an entire hotel under the bus to save your story, or can you muster the courage to update your position with dignity?

ABOVE ALL, AN HONORABLE SALESMAN IS A SAPIENT SALESMAN.